Lecture Tutorials
for Introductory Geoscience

Karen M. Kortz
Jessica J. Smay

W. H. Freeman and Company
New York

Printed in the United States of America

ISBN-10: 1-4292-5378-9

ISBN-13: 978-1-4292-5378-9

First printing

W. H. Freeman and Company
41 Madison Avenue
New York, NY 10010
Houndsmills, Basingstoke
RG21 6XS England
www.whfreeman.com

Table of Contents

What are Lecture Tutorials?[1]

Each Lecture Tutorial is a short worksheet that students complete in class, making the lecture more interactive. Research indicates that students learn more when they are actively engaged while learning, and several studies indicate that students who use Lecture Tutorials in the classroom retain more knowledge than students who only listen to a lecture on the same material. After a brief lecture on the subject, students work in small groups to complete the Lecture Tutorial worksheets.

The Lecture Tutorials are designed to address misconceptions and other topics with which students have difficulties. They create an environment where students confront their misconceptions and, through well-designed questioning, guide students to a more scientific way of thinking. This careful design makes Lecture Tutorials unique among most other activities used in the classroom.

By posing questions of increasing conceptual difficulty to the students, Lecture Tutorials help students construct correct scientific ideas. The first questions help the students think about what they do and do not know. The Lecture Tutorial then guides the students by asking them questions focused on underdeveloped or misunderstood concepts and slowly steps them through thinking about more difficult questions. The final questions on the Lecture Tutorial help to indicate whether the students understand the material.

Lecture Tutorials can be used in any size classroom. Students should speak with each other and teach each other while the instructor acts as a facilitator. The questions on the Lecture Tutorials require no technology and are written so the conceptual steps for each question are manageable.

Guidelines for Use

Step-By-Step Implementation for the Instructor

1. Lecture on the material as usual. You can also provide an introduction of the background information that students need to know before beginning the Lecture Tutorial.
2. Optional: Pose a well-designed, multiple-choice question for you and the students to gauge their understanding of the material.
3. Have the students split into groups of 2 or 3 and work on the Lecture Tutorial. Walk around the room and answer their questions. Lecture Tutorials take 10–15 minutes for most students to complete.
4. Review some of the main points of the Lecture Tutorial.
5. Optional: Pose a new multiple-choice question to check if the students have the expected understanding of the information.
6. Continue with the lecture.

Directions for the Student

You are using Lecture Tutorials in your class because they help to improve your understanding of the material. They require you to actively think through questions instead of listening passively to the lecture. Lecture Tutorials allow you to gauge how well you understand the

[1] Revised from On the Cutting Edge – Teaching Methods – Lecture Tutorials (http://serc.carleton.edu/NAGTWorkshops/teaching_methods/lecture_tutorials/index.html).

material and ask any necessary questions. They also address different learning styles, so you can use your strengths while learning the material. Surveys have shown that an overwhelming majority of students feel that Lecture Tutorials are a useful part of their learning experience.

However, in order for you to get maximum benefits from the Lecture Tutorials, you need to put effort into completing them. Think about the answers as you are working through them, and be sure to write down your logic. Nothing is more frustrating than reviewing the Lecture Tutorials and not remembering how you solved the problems! You will be asked to work in groups to complete the Lecture Tutorials. Take advantage of working with your fellow students by both learning from them and teaching them.

Acknowledgements

We wish to credit Scott Clark for his original research on misconceptions addressed in parts of five Lecture Tutorials (Tectonic Plates and Boundaries, Subduction Features, Movement at Convergent Boundaries, Plate Boundaries in Oceans, Melting Rocks and Plate Tectonics). Many of the figures and lines of questioning in these Lecture Tutorials are adapted from his unpublished research instruments. Initial findings were presented in Clark, S.K., Libarkin, J.C. (2008) Post-Instruction Alternative Conceptions about Plate Tectonics Held by Non-Science Majors (Abstract 248-14 at the Geological Society of America Meeting in 2008).

We would like to thank Ed Prather and Timothy F. Slater for their guidance and encouragement. We learned of the Lecture Tutorial teaching method through the Lecture Tutorials for Introductory Astronomy they co-wrote and promoted, and after discussions with them, we created our own geoscience Lecture Tutorials.

We appreciate the enthusiasm and hard work done by Anthony Palmiotto, Anthony Petrites, Jodi Isman, Christine Buese, Amy Thorne, Scott Guile, Randi Rossignol, Clancy Marshall, and the rest of the team at W. H. Freeman.

Each of these Lecture Tutorials has gone though many revisions, and we would like to thank all of the students who gave us feedback and comments on the previous versions of these Lecture Tutorials.

We also wish to thank Brian and Greg. Their understanding, advice, and support made this work possible.

Partial support for this work was provided by the National Science Foundation's Course, Curriculum, and Laboratory Improvement (CCLI) program under Award No. 0837185. Any opinions, findings, and conclusions or recommendations expressed in this material are those of the authors and do not necessarily reflect the views of the National Science Foundation.

Below is a map showing tectonic plate boundaries. The edges, or boundaries, of the tectonic plates are labeled.

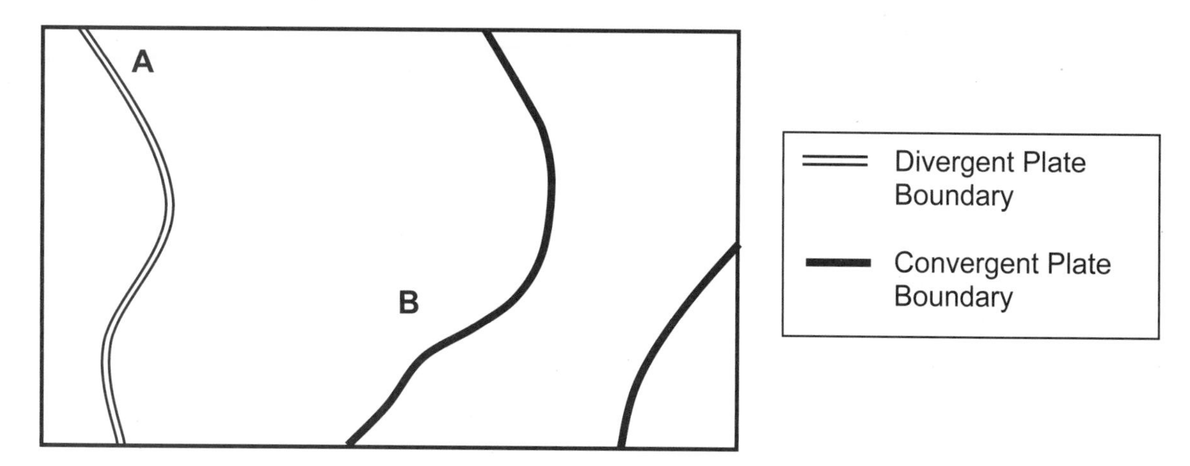

1) How many individual tectonic plate boundaries are in the diagram? 1 2 3 4 5 6 7

2) How many tectonic plates are in the diagram? 1 2 3 4 5 6 7

3) Are A and B on the same tectonic plate? Yes No

Below is a map of the same tectonic plate boundaries. However, this map also shows the location of ocean and land.

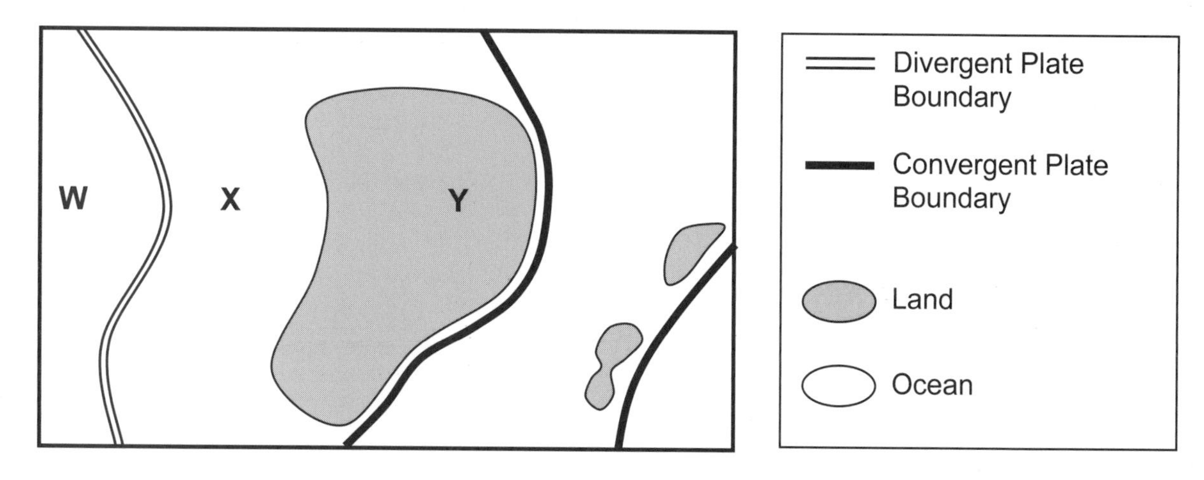

4) How many tectonic plates are in the diagram? 1 2 3 4 5 6 7

5) Why should your answers to Questions 2 and 4 match?

6) Are W and X on the same tectonic plate? Yes No

7) Are X and Y on the same tectonic plate? Yes No

8) Three students are discussing the tectonic plate on which X is located.

Student 1: *X and Y are both on the same plate because there is no boundary between them, but X is on a different plate than W because there is a divergent boundary between them.*

Student 2: *X and Y are on different plates because X is in the ocean and Y is on a continent. X is on the same plate as W because they are both in the ocean.*

Student 3: *X is on a different plate than both Y and W. There is a divergent plate boundary separating X and W and an edge of a continent separating X and Y.*

With which student do you agree? Why?

9) How is a tectonic plate boundary different than a tectonic plate?

10) Explain why a continent is different than a tectonic plate.

Part 1: Divergent Boundary

The diagram below is of a divergent boundary with arrows showing the direction in which the plates are moving.

A B C

The scale lines are 10 km apart

1) Where is the oldest crust found?

A B C

2) What is the age of the rocks at location B?

0 years old 1 million years old 2 million years old 4 million years old

3) If each plate is moving at a rate of 20 km per 1 million years, roughly how long did it take for Rock A to reach its current location?

0 years 1 million years 2 million years 4 million years

4) What is the age of the rock at location C?

0 years old 1 million years old 2 million years old 4 million years old

5) Why should your answer to Question 4 be twice your answer to Question 3? Revise your answers if necessary.

6) A map of the Atlantic Ocean is shown to the right. Where are the oldest rocks in the Atlantic found?

D E

Briefly explain your answer.

7) Two students are debating about the relative ages of the rocks that make up the crust in the Atlantic Ocean.

Student 1: *The oldest rocks are located at E because it is the farthest from a continent. The rocks would take a really long time to get to the middle of the ocean.*

Student 2: *But this ocean has a divergent boundary in the center. This means that rocks at E are really young. D is farthest from the divergent boundary, so that's where the oldest rocks are.*

With which student do you agree? Why?

Part 2: The Atlantic Ocean

Examine the map of the ages of the seafloor in the Atlantic Ocean.

8) Does the pattern of ages match your answer to Question 6? Revise your answer if necessary.

9) Draw a line along the divergent boundary.

10) What is the age of the oldest rocks in the Atlantic Ocean?

11) Approximately how long ago did the Atlantic Ocean begin to form?

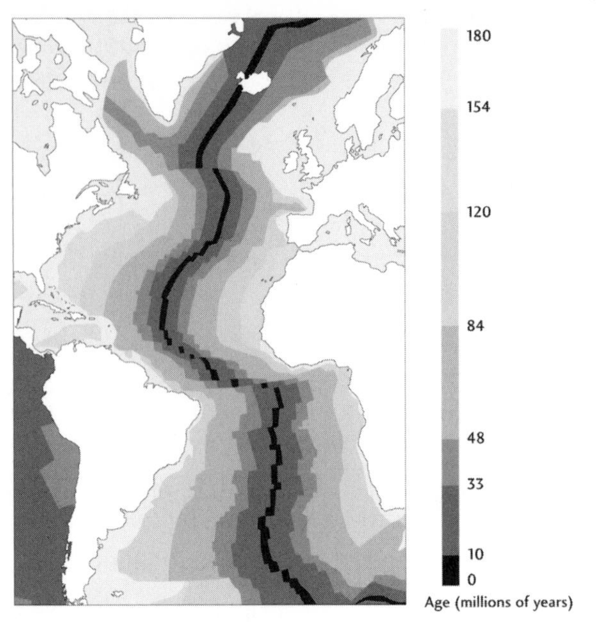

Map of the ages of the seafloor in the Atlantic Ocean

12) Why should your answers to Questions 10 and 11 match? Revise your answers if necessary.

13) You are reading a proposal requesting money to search for evidence of a crater that caused a mass extinction on Earth around 250 million years ago. The team is proposing to search a poorly explored area of the floor of the Atlantic Ocean between South America and northern Africa. Would you fund this project? Use the ages of the seafloor to support your answer.

Compare your answer to the last question with the answers of other groups.

Part 1: Features at Convergent Plate Boundaries with Subduction

The cross section below shows a subduction zone at an ocean-continent convergent boundary. The ocean surface is indicated by a dashed line.

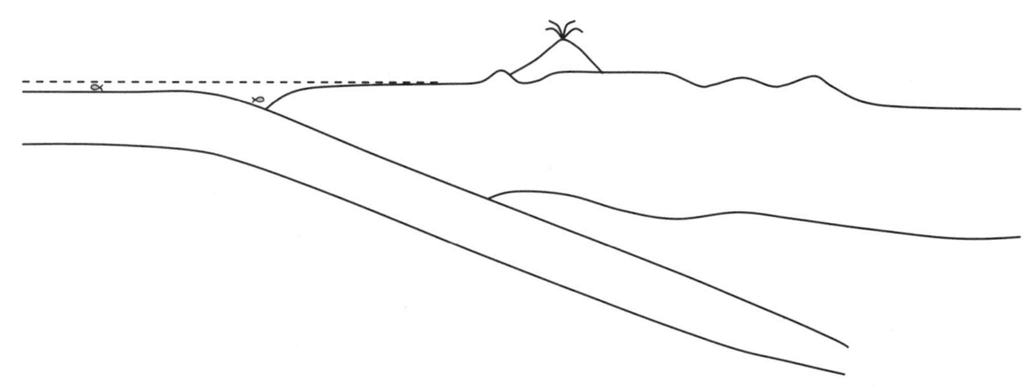

1) Draw two arrows on each plate showing which way the plates are moving relative to each other.

2) On the diagram, label features that geologists could use to identify this plate boundary.

3) For the features listed below, briefly explain how they formed.

   Volcanoes and mountains:

   Ocean trench:

4) Subduction of an ocean plate takes many millions of years. If you were examining a map, what would you look for to indicate that subduction is happening, even if you cannot watch?

5) Sketch one scenario that might occur when two ocean plates move toward each other. Label the trench and volcanoes; the type of plate has been labeled for you.

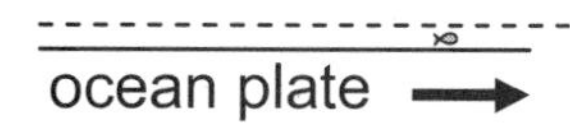

← ocean plate

Subduction Features

Part 2: Plate Boundary Location

The cross section below shows a subduction zone at an ocean-continent convergent boundary. The ocean surface is indicated by a dashed line.

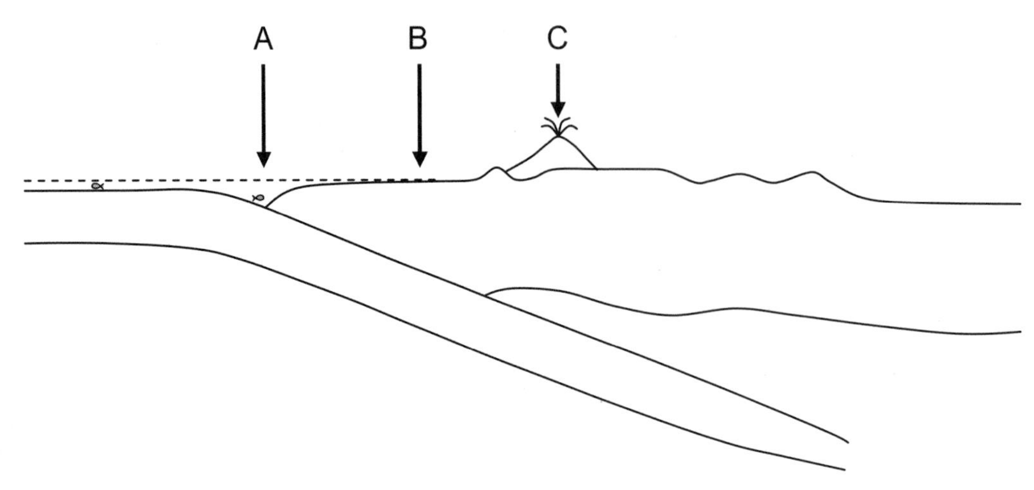

6) On the diagram, label each of the following features at the corresponding arrow:
   - Volcanoes and mountains
   - Coastline
   - Ocean trench

7) Three students are discussing which arrow points to the location on the surface of the convergent plate boundary.

Student 1: *I think that Arrow A points to the plate boundary because that is where one plate meets the other on the diagram.*

Student 2: *I think that Arrow B points to the plate boundary because that is where the ocean turns into continent.*

Student 3: *I think that Arrow C points to the plate boundary because that is where the geologic action is happening.*

With which student do you agree? Why?

8) You are given a map of an area with a subduction zone. Explain what feature you would use on the map to determine the exact location of the plate boundary.

Movement at Convergent Boundaries

Part 1: Movement Shown on a Cross Section

The cross sections below show subduction zones at an ocean-continent convergent plate boundary. We will examine the boundary relative to the location of the trench. In other words, the general location of the trench does not change, but other things might move around it.

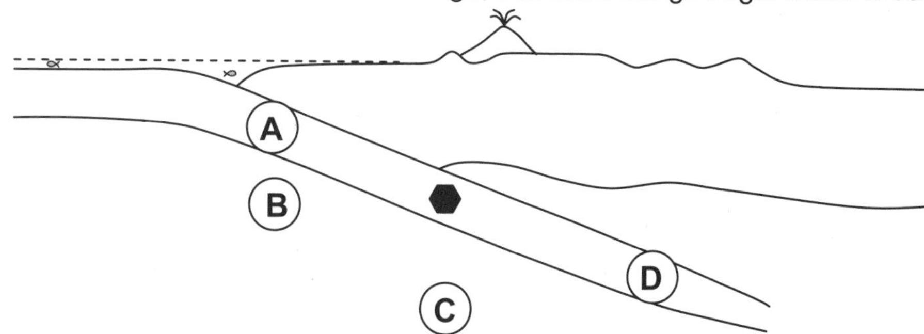

1) Draw two arrows on each plate indicating the relative direction the plates are moving.

2) Where was ⬢ in the past? A B C D same place

3) Where will ⬢ be in the future? A B C D same place

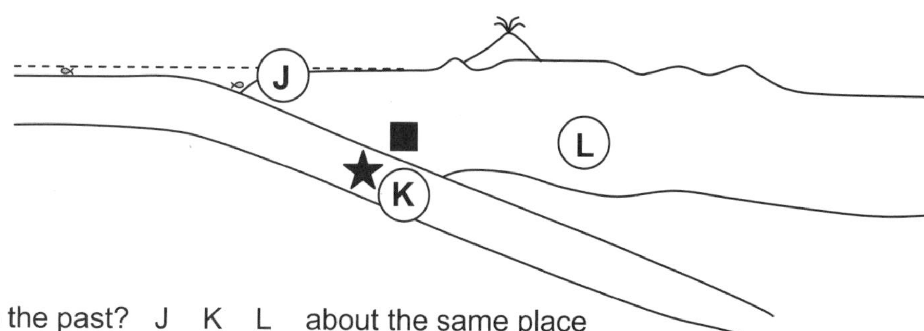

4) Where was ■ in the past? J K L about the same place

5) Where will ■ be in the future? J K L about the same place

6) Two students are discussing how the ■ on the continental plate will move over time relative to the trench.

Student 1: *I think that the square will stay in about the same place relative to the trench because the ocean plate is the plate that is subducting and is destroyed. The continental plate is scrunched a little, but it isn't destroyed like the ocean plate is.*

Student 2: *But it's a convergent boundary, and the plates are moving together. Because I can draw arrows showing the plates moving together, that means that the square is moving toward the ocean plate, away from the volcano and closer to the trench.*

With which student do you agree? Why?

7) In 50 million years, will ■ and ★ be relatively close to each other, as they are now? yes no
Explain your answer.

Movement at Convergent Boundaries

Part 2: Movement Shown on a Map View

The map below shows the surface features at a subduction zone at an ocean-continent convergent plate boundary. We will examine the movement of features relative to the location of the trench. In other words, the general location of the trench does not change, but other things might move around it.

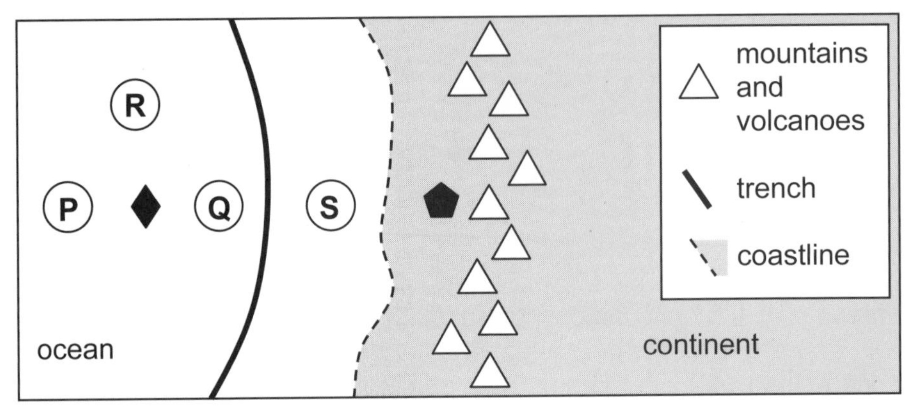

8) Draw a line along the plate boundary.

9) Draw two arrows on each plate indicating the relative direction the plates are moving.

10) Where was ◆ in the past? P Q R S same place

11) Where will ◆ be in the future? P Q R S same place

12) At what location will ◆ disappear from view? trench coastline mountains/volcanoes

13) Assuming that both ◆ and ⬟ are locations on the surface of Earth, will they ever meet? Explain your answer.

Part 1: Features

The cross section below shows the tectonic plates beneath an ocean and a nearby continent. Sea level is indicated by the dashed line.

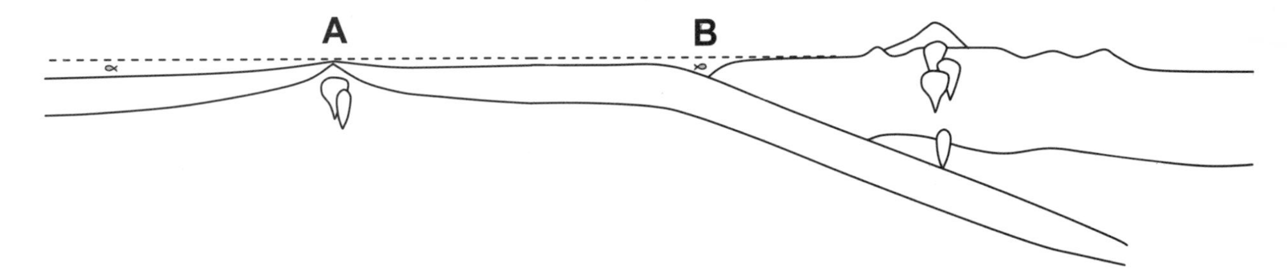

1) Use arrows to show the directions the plates are moving near Locations A and B.

2) Two students are discussing the directions the plates are moving.

Student 1: *I drew the arrows moving apart at Location A because magma is pushing the plates up as it rises to fill in the gap as the plates move apart.*

Student 2: *I drew the arrows moving together at Location A because as the plates are moving toward each other, they get pushed up.*

With which student do you agree? Why?

3) What seafloor feature is found at Location A? ridge trench abyssal plain island

4) What seafloor feature is found at Location B? ridge trench abyssal plain island

5) What type of plate boundary is at A? divergent convergent transform

6) What type of plate boundary is at B? divergent convergent transform

7) Check that your answers for Questions 5 and 6 match the arrows you drew at Locations A and B in Question 1.

8) If you find a plate boundary in the middle of an ocean away from the edge, what type of plate boundary is it most likely to be?

divergent convergent with subduction convergent without subduction

9) If you find a plate boundary along the edge of an ocean next to a continent, what type of plate boundary is it most likely to be?

divergent convergent with subduction convergent without subduction

Plate Boundaries in Oceans

Below is a cross section of the ocean floor and nearby land showing the surface features. Sea level is indicated by the dashed line.

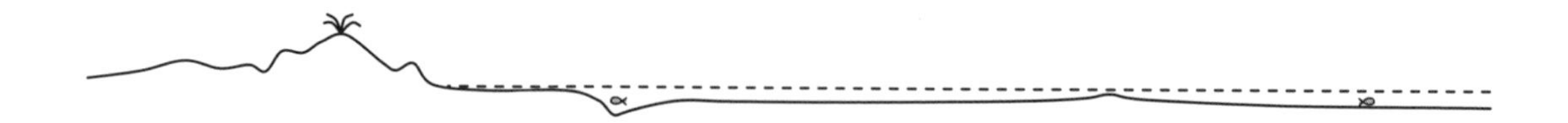

10) Label the ocean ridge.

11) Label the ocean trench.

12) Label the mountains/volcanoes on land.

13) Label the divergent boundary.

14) Label the convergent boundary.

15) Draw what the plates are doing beneath the surface to produce the surface features.

Below are cross sections of an ocean at different stages of its life cycle, from pre-ocean to post-ocean. The dotted crust is ocean crust, and the solid white crust is continental crust. The dashed line indicates the ocean surface (sea level).

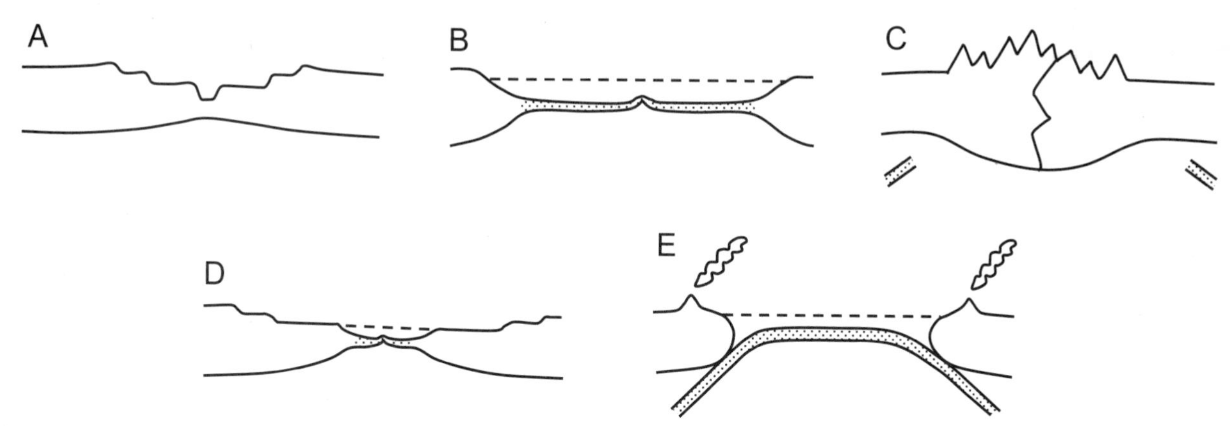

1) Describe the one plate boundary shown in B. Is the ocean growing or shrinking?

2) Describe the two plate boundaries shown in E. Is the ocean growing or shrinking?

3) Arrange the five diagrams in sequential order from pre-ocean to post-ocean. Write your order below.

Youngest ________ ________ ________ ________ ________ Oldest

4) Two students are debating the order of the stages of an ocean.

Student 1: *I put the cross sections in order from the smallest ocean to the biggest: C-A-D-E-B. I did this because the ocean is growing bigger over time.*

Student 2: *But you need to look at the plate boundaries for clues. E has subduction happening, so that means the ocean is shrinking, and that means it was once bigger.*

With which student do you agree? Why?

History of an Ocean

The way that people typically think about the age of something is to take that one thing and look at it as it changes over time (e.g., pictures of you as a baby, toddler, child, and adult). However, this often does not work in geology because it can take millions of years for something to change. Oceans are one example of this. So, another way to look at the stages of an ocean is to look at all the oceans today and put them in order of how old they are (e.g., compare a toddler in one city to a teen in another city).

5) Next, match the diagrams from the previous page, which illustrate the stages of an ocean, to the five locations listed below. Place the letter next to the ocean.

_____ East African Rift

_____ Atlantic Ocean

_____ Himalayas

_____ Red Sea

_____ Pacific Ocean

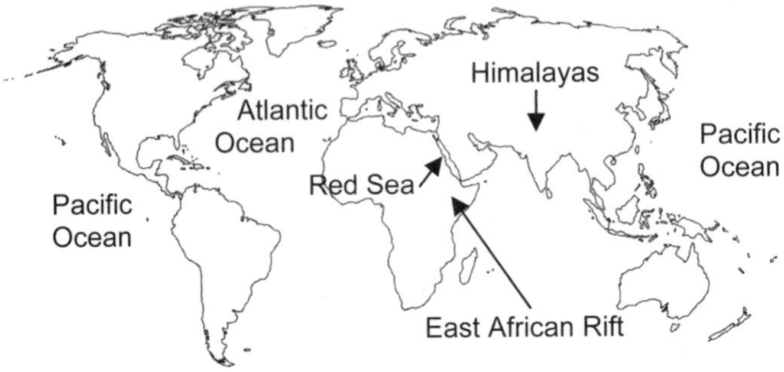

6) Take your answers to Questions 1 and 5 and place the locations on Earth in order from the youngest to the oldest in the life cycle of an ocean.

Youngest ____________________ ____________________

____________________ ____________________ ____________________ Oldest

7) Two students are debating the locations on Earth and their order in the stages of an ocean.

Student 1: *I think the Himalayas are the very end of an ocean. They formed as a result of a convergent boundary, so the plates are coming together, which means the ocean that was once there completely shrunk away into nothing*

Student 2: *No. I think the Himalayas are the very beginning of an ocean because they are the highest, and then the crust would get thinner and thinner until it becomes an ocean, and the ocean would then grow bigger.*

With which student do you agree? Why?

8) Revise the order of your ocean stages in Questions 3 and 6 if necessary.

9) Predict what the Red Sea will look like in 100 million years.

10) Predict what the East Coast of the United States could look like in 100 million years.

Features on the Ocean Floor

The ocean floor is primarily oceanic crust, but also includes the edges of the continent (the continental margins). Although once thought to be featureless plains, there are many features that have been discovered on the ocean floor.

1) On the diagram below, draw arrows and label the ocean floor features:

Two continental shelves *Abyssal plain (choose one)* *Ocean trench*

Island *Ocean ridge*

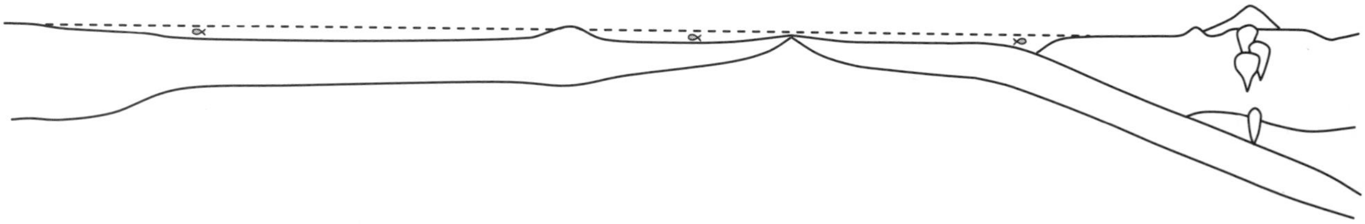

Use the information from the diagram to answer the following questions about different features found on the ocean floor.

2) Which landform on the ocean floor is related to convergent plate boundaries?

____

Explain.

3) Which landform on the ocean floor is related to divergent plate boundaries?

____

Explain.

4) Which landforms on the ocean floor can form independent of plate boundaries?

______________________________ ______________________________ ______________________________

Explain.

Features on the Ocean Floor

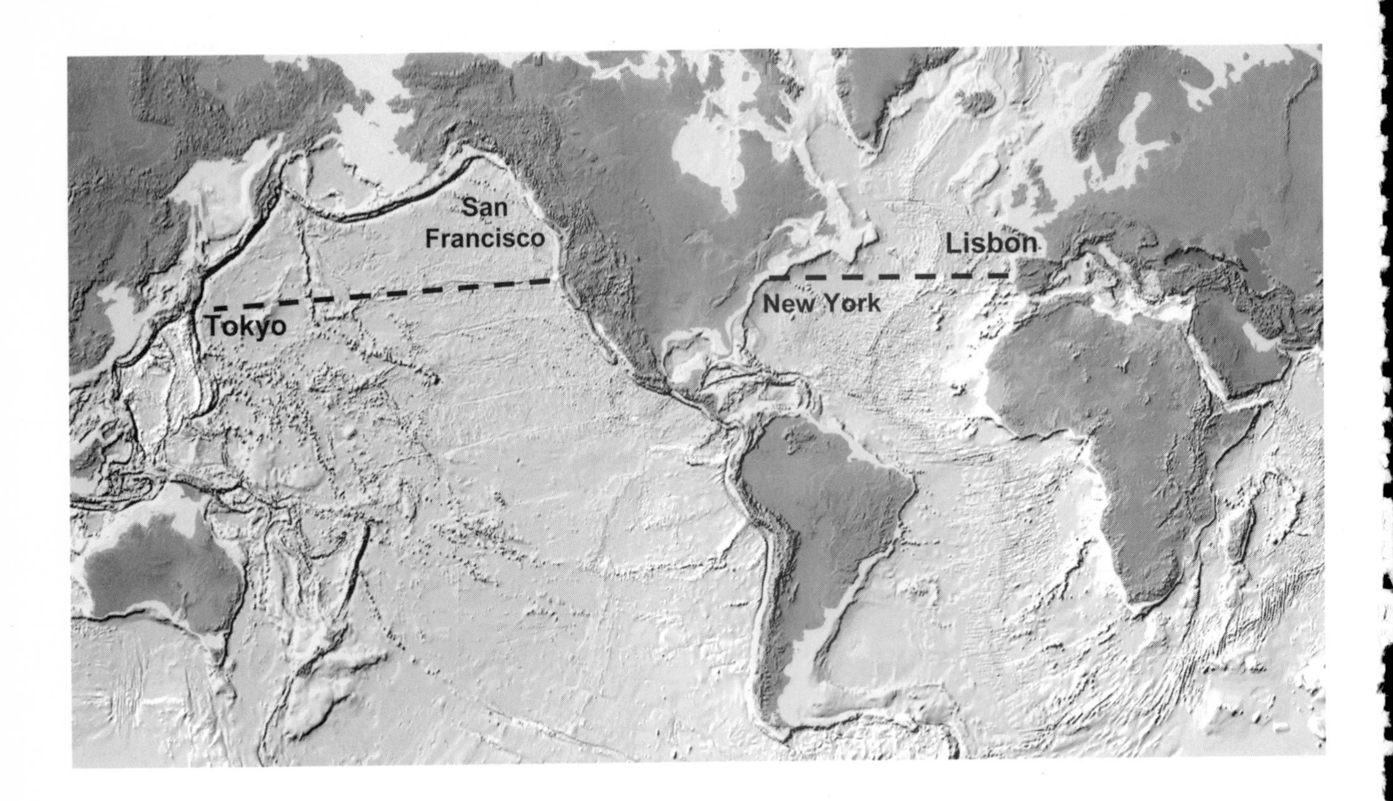

5) If you could walk along the Atlantic Ocean seafloor from New York to Lisbon, Portugal, what landforms would you walk past?

6) If you could walk along the Pacific Ocean seafloor from San Francisco to Tokyo, Japan, what landforms would you walk past?

7) Explain how the landforms on the ocean floor can be used to determine where plate boundaries are located. Give examples to support your answer.

Part 1: Convergent Boundaries

Below is a cross section showing subduction of an ocean plate beneath a continental plate. The ocean surface is shown by the dashed line.

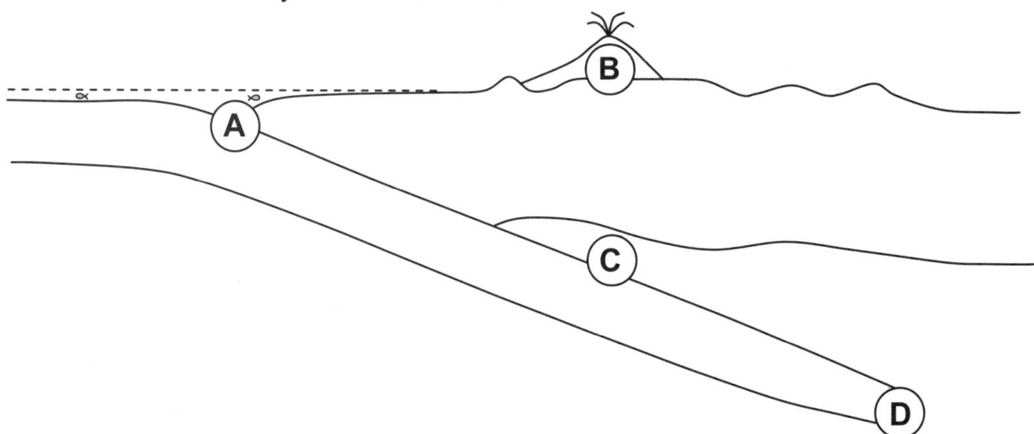

1) Where are rocks melting to produce magma for the volcanoes?

A B C D

2) Four students are discussing where the rocks are melting.

Student 1: *I picked A because that's where the ocean plate is being subducted beneath the continental plate.*

Student 2: *I picked B because that's where the volcanoes are, so that's where the magma is.*

Student 3: *I picked C because the volcanoes are right above it, and it's deep enough to be hot enough to melt.*

Student 4: *I picked D because that's where the subducting ocean plate disappears in the diagram, and it's deep enough to be hot enough to melt.*

With which student do you agree? Why?

Rocks can melt when either the temperature increases, the pressure decreases, or water is added to the rock. As you go deeper beneath Earth's surface, the temperature increases. As you rise to Earth's surface, the pressure decreases.

3) Rocks cannot melt in the top of the crust because it is not hot enough. Therefore, in what two locations could the rocks melt?

A B C D

4) Volcanoes form directly above where rocks melt. Magma is less dense than the surrounding rock, so it rises straight through the crust forming magma chambers for volcanoes. Where could rocks melt to form magma that rises into magma chambers and erupts from volcanoes?

A B C D

5) As the ocean plate subducts and the rocks get hot deep beneath the surface, the minerals in the rocks lose water, adding water to the surrounding rock. Therefore, where could water be added to the rock causing it to melt?

A B C D

6) Based on your answers to Questions 4, 5, and 6, where do rocks melt to produce magma for the volcanoes?

A B C D

7) Draw an arrow on the diagram to show how the magma moves straight up from its melting location to erupt out of the volcano.

8) Summarize why rocks do not melt to produce magma at the other three locations at the convergent boundary.

Part 2: Divergent Plate Boundaries

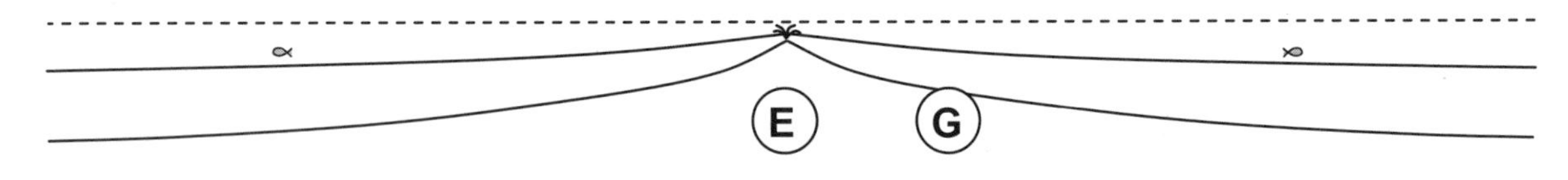

9) Why do rocks melt at E but not at F?

hotter temperature lower pressure more water

10) Why do rocks melt at E but not at G?

hotter temperature lower pressure more water

Part 3: Hotspots

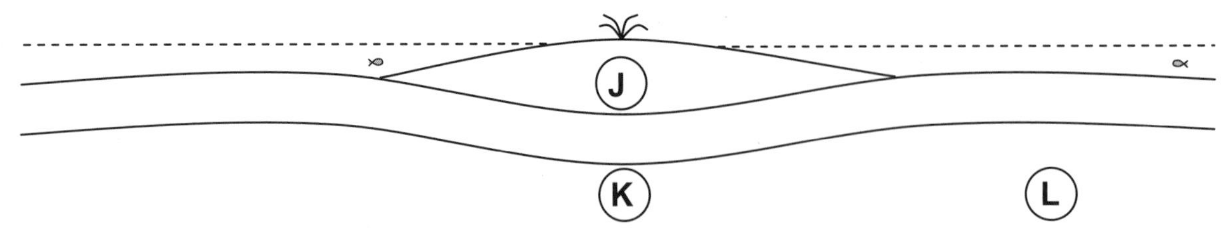

11) Why do rocks melt at K but not at J?

hotter temperature lower pressure more water

12) Why do rocks melt at K but not at L?

hotter temperature lower pressure more water

Part 4: Putting It Together

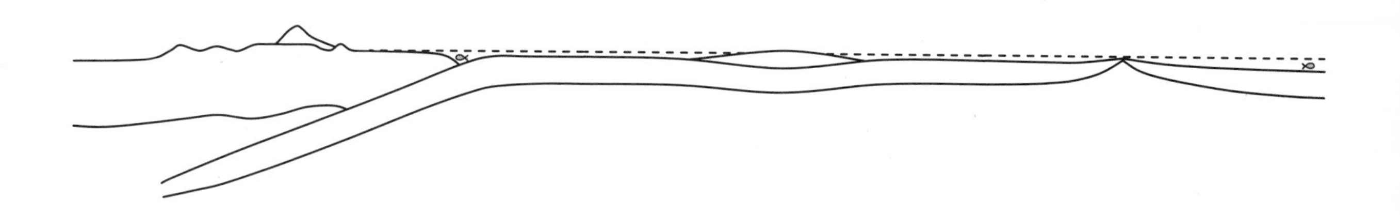

13) On the cross section above, put an "X" exactly at the three locations where rocks melt to form magma that will move up and erupt at the surface.

14) Explain why rocks melt in these three locations:

Melting location 1:

Melting location 2:

Melting location 3:

15) On the cross section above, draw a star at each of the three places on the surface where lava erupts.

16) Draw an arrow connecting each melting location you labeled with the X with the place that the lava erupts at the surface (labeled with a star).

17) Compare your answers to Questions 13–16 with the work you did earlier in the worksheet; be sure your answers agree.

Part 1: Determining the Size of Earth's Outer Core

P and S seismic (earthquake) waves are used to determine the composition and phase of the interior of Earth. The diagram below represents Earth; the star is the location of an earthquake, and the tick marks indicate seismic stations that measure P and S waves.

P waves arrive first and do travel through liquid.
S waves arrive second and do not travel through liquid.
Surface waves arrive last and are the largest.

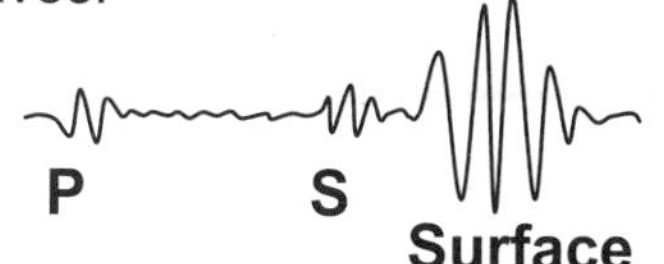

1) Circle the **S** waves on each of the seismic stations that recorded **S** waves. Some have been done for you.

2) Draw arrows representing the path of **S** waves from the earthquake to each seismic station that recorded **S** waves. Only draw lines for **S** waves. Some have been done for you.

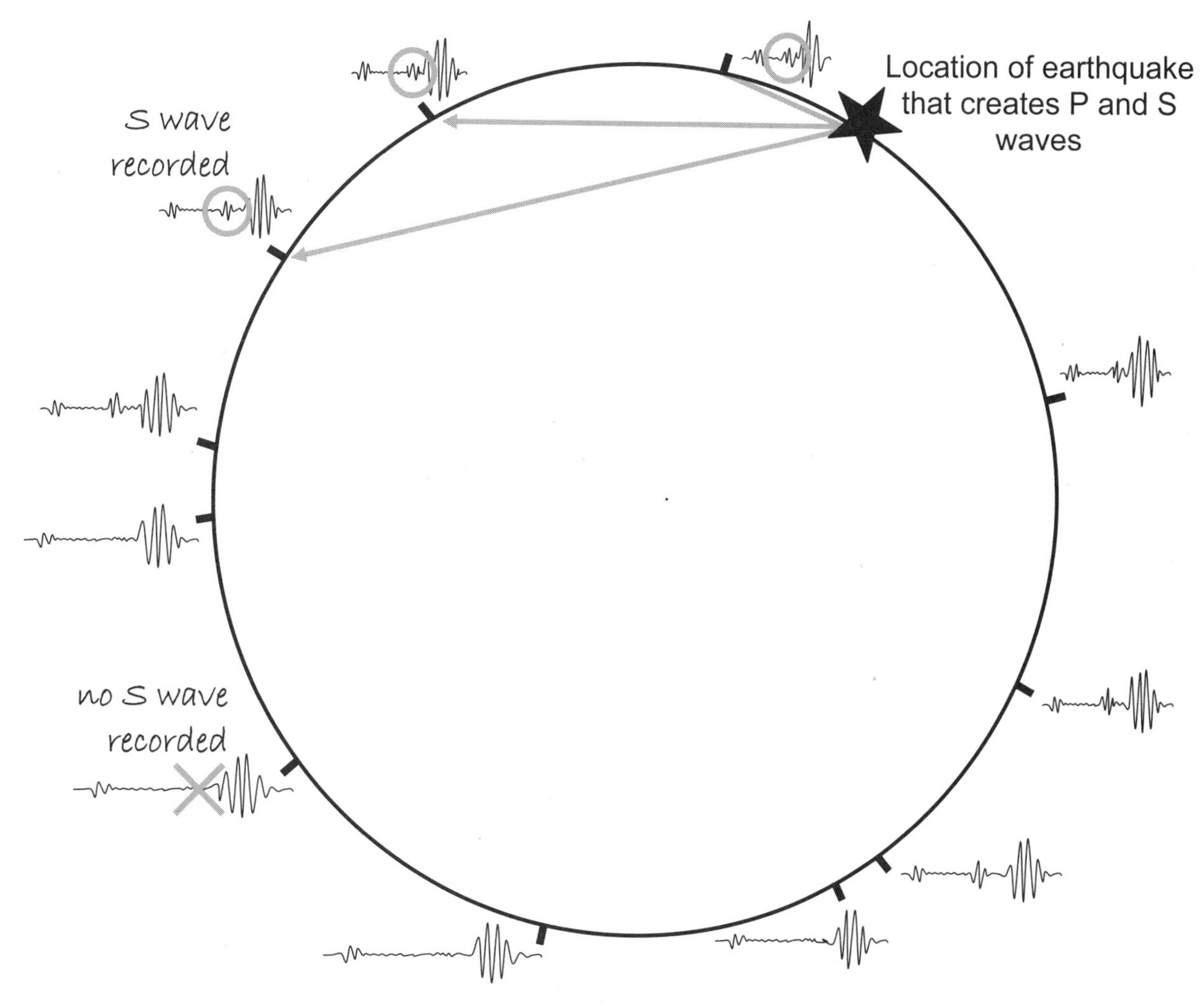

3) Based on the pattern of S waves, what can you determine about the phase of the outer core of Earth? Is it solid or liquid?

4) Draw the outer core on the diagram, using the distribution of S waves to help you determine the size.

The Outer Core

5) The mantle is the layer between the outer core and the crust at the surface of Earth. Based on where S waves are detected around Earth, how much of the mantle is liquid?

very little to none   about half   most to all

Explain your answer.

Part 2: The Outer Core of Other Planets

6) Does this planet have a large or small molten core?

P = arrival of P waves
S = arrival of S waves

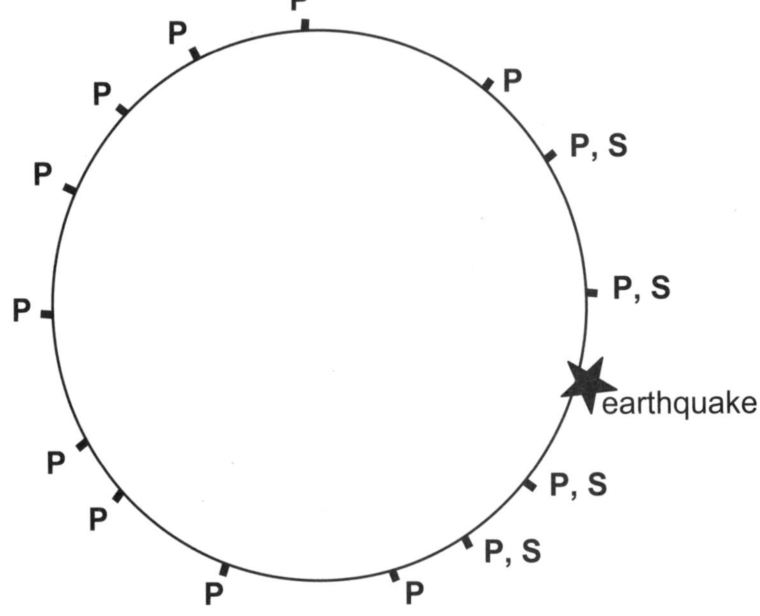

7) What can you conclude about the interior of this planet?

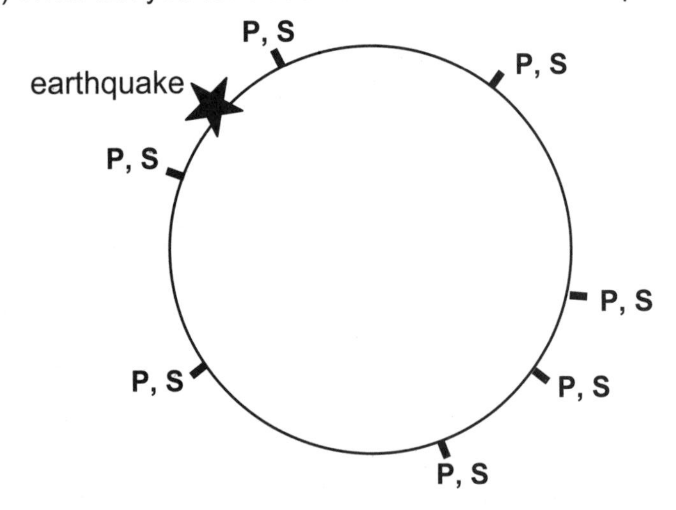

Part 1: Earth's Layers

Earth can be divided into three layers based on composition (what it is made of): the crust, mantle, and core. The core can be divided into two layers based on phase: the liquid outer core and the solid inner core. The depths in the chart below are approximate and vary with location.

Layer	Depth of top	Depth of bottom	Phase and composition
Crust	surface of Earth	30 km	solid, rock
Mantle	30 km	2900 km	mostly solid, rock
Outer core	2900 km	5100 km	liquid, metal
Inner core	5100 km	center of Earth	solid, metal

1) Sketch and label the four layers of the Earth on the diagram below. The inner core has been drawn and labeled for you.

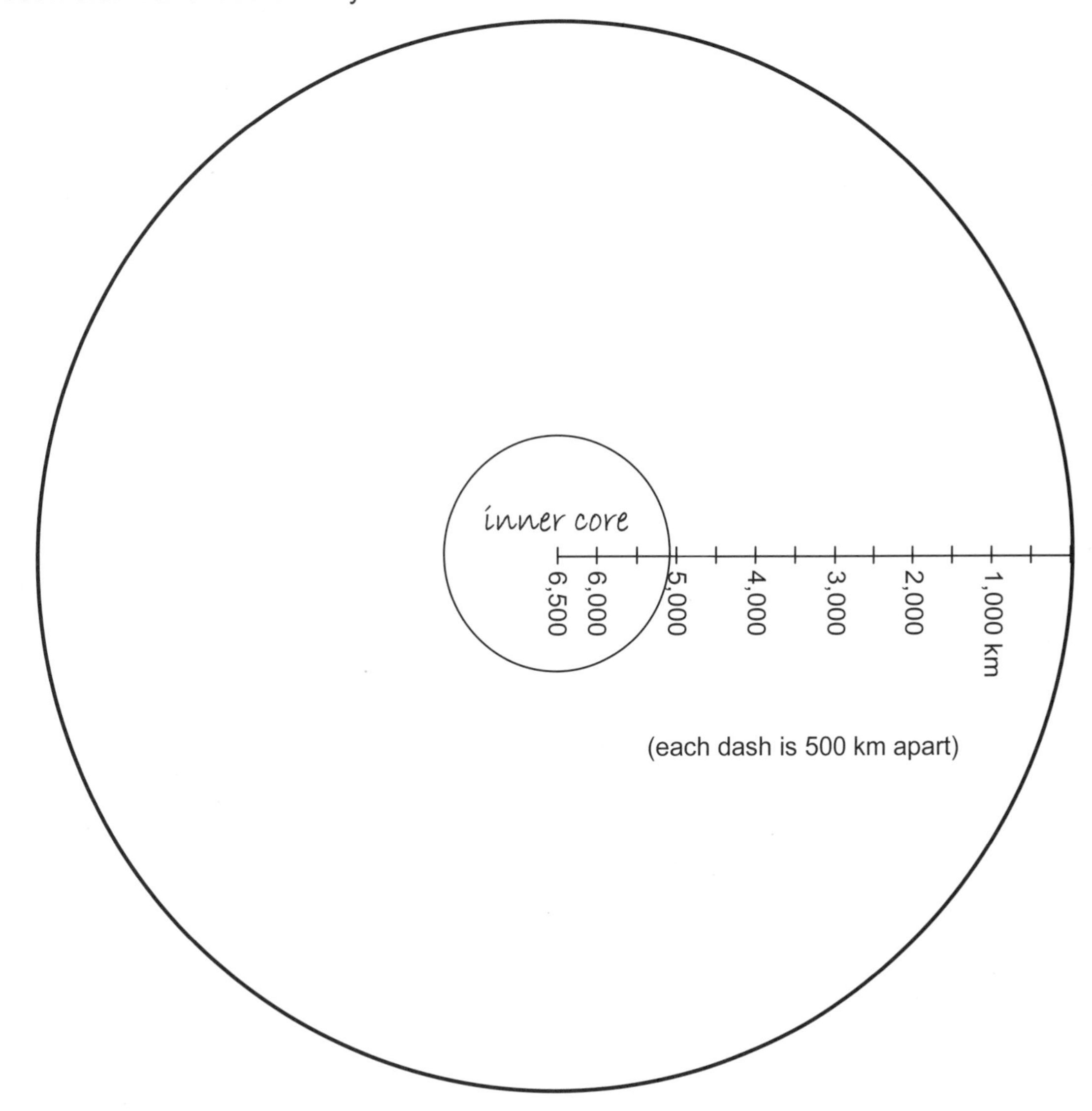

2) What is the best comparison for the thickness of the crust?
   a. The crust has the same relative thickness as the skin of an apple.
   b. The crust has the same relative thickness as the peel of an orange.

Part 2: Origins of Magma

Materials melt with a combination of high temperature and low pressure. If the temperature is too low or the pressure is too high, rocks will not melt.

The outer core has the right combination of temperature and pressure for metal to be molten. At 15 to 100 km below the surface, the temperature and pressure are potentially just right to partially melt rock. This depth is where pockets of most volcanic magma are formed.

3) On the diagram of Earth, draw a star at the depth of the source of magma.

4) What layers melt to form magma?

Crust Upper (outer) mantle Lower (inner) mantle Outer core Inner core

5) Overall, how much of the mantle is liquid? very little to none about half most to all

6) According to your diagram, estimate how far the molten metal from the outer core would have to travel to erupt as a volcano. For comparison, New York and Los Angeles are ~4000 km apart.

7) What is the composition of the outer core? What is the composition of erupted lava from a volcano?

8) Two students are debating whether the molten outer core erupts as volcanoes.

Student 1: *I don't think the molten outer core erupts as volcanoes because the magma would have to travel thousands of kilometers through the mantle to reach the surface, and I don't think it could go that far through the mostly solid mantle.*

Student 2: *If the outer core erupted as volcanoes, then we would have pure metal erupting out of Earth's surface. Volcanoes erupt molten rock, so the molten source cannot be the outer core.*

Do you agree with one or both students? Why?

9) You are the science advisor to a movie. The screenwriters come to you with the following scenario: A mad scientist threatens to detonate a bomb in the center of Earth, triggering volcanoes around the world to erupt, unless world leaders pay him a large ransom. Explain to the screenwriters why their story is or is not scientifically accurate. (*Note: Atomic bombs at Earth's surface can cause damage up to 20 km away.*)

1) Examine the minerals and their corresponding chemical formulas (see chemical symbols below). Circle the key aspects of the chemical formula and determine which mineral group each mineral falls into (use the groups listed below).

Chemical Symbols	
C	Carbon
Fe	Iron
Mg	Magnesium
O	Oxygen
Si	Silicon

Groups of Minerals → Description

- Silicates (has silicon and oxygen)
  - Non-Ferromagnesian Silicates (no iron, no magnesium) → Light color, not metallic
  - Ferromagnesian Silicates (also has iron and magnesium) → Dark color, not metallic
- Carbonates (has carbon and oxygen: CO_3) → Often fizzes w/acid
- Other (no silicon, often a metal, salt or sulfate) → Sometimes metallic

Mineral name, chemical formula, and description	Key aspects of the chemical formula (circle or cross out)	Mineral group (circle one)
Quartz $Si\,O_2$ Light colored, not metallic	Si C O Fe Mg	Non-Fe Mg silicate Fe Mg silicate Carbonate Other
K-feldspar $K\,Al\,Si_3\,O_8$ Light colored, not metallic	Si C O Fe Mg	Non-Fe Mg silicate Fe Mg silicate Carbonate Other
Calcite $Ca\,C\,O_3$ Light colored and reacts with acid, not metallic	Si C O Fe Mg	Non-Fe Mg silicate Fe Mg silicate Carbonate Other
Biotite $K\,(Mg,Fe)_3\,Al\,Si_3\,O_{10}\,(O\,H)_2$ Dark colored, not metallic	Si C O Fe Mg	Non-Fe Mg silicate Fe Mg silicate Carbonate Other
Pyrite $Fe\,S_2$ metallic	Si C O Fe Mg	Non-Fe Mg silicate Fe Mg silicate Carbonate Other
Halite (Salt) Na Cl Light colored, not metallic	Si C O Fe Mg	Non-Fe Mg silicate Fe Mg silicate Carbonate Other

2) Two students are debating the classification of graphite (chemical formula: C).

Student 1: *I think that graphite belongs in the "Other" category. It has carbon, but it doesn't have oxygen, so it's not a carbonate.*

Student 2: *But the word "carbonate" has the word "carbon" in it! I think that if any mineral has carbon in its chemical formula, it must be a carbonate.*

With which student do you agree? Why?

3) Pencil lead is made of graphite. What about the appearance of graphite indicates that it is in the "Other" mineral group?

4) What is the classification of hematite (Fe_2O_3)? Explain your answer using the key elements in the chemical formula.

5) Muscovite mica ($KAl_2AlSi_3O_{10}(OH)_2$) has nearly the same chemical formula as biotite mica ($K(Mg,Fe)_3AlSi_3O_{10}(OH)_2$). They both form crystals that are platy sheets. Based on their chemical formula, predict how they are different in appearance.

6) The igneous rock gabbro is made up primarily of ferromagnesian silicate minerals. Predict what gabbro looks like. Explain your answer.

Part 1: The Rock Types

The three types of rocks are igneous, sedimentary, and metamorphic rocks.

1) Briefly describe how each of the rock types form.

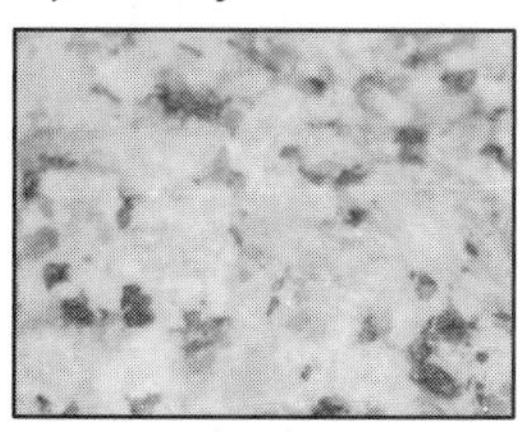

Igneous:

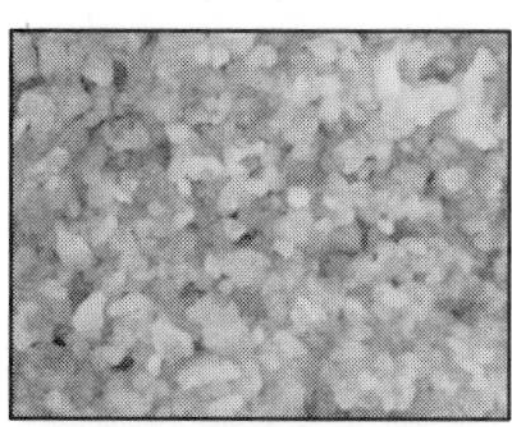

Sedimentary:

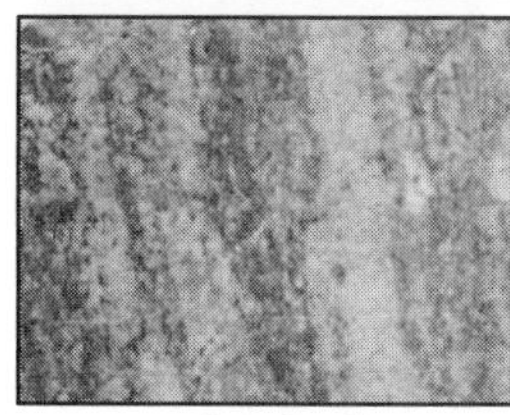

Metamorphic:

Jessica Smay

2) Can igneous rocks form from the following rock types?

Igneous Y or N Sedimentary Y or N Metamorphic Y or N

3) Can sedimentary rocks form from the following rock types?

Igneous Y or N Sedimentary Y or N Metamorphic Y or N

4) Can metamorphic rocks form from the following rock types?

Igneous Y or N Sedimentary Y or N Metamorphic Y or N

5) Two students are debating the answers to Questions 2–4.

Student 1: *I think that the different rock types can form from the other rock types, but they can't form from themselves. For example, igneous rocks can form from sedimentary and metamorphic rocks, but not other igneous rocks.*

Student 2: *Why not? If metamorphic rocks and sedimentary rocks can melt and form igneous rocks, why can't igneous rocks melt and form igneous rocks again? I think that all rock types can form from all other rock types. So I circled Y for everything.*

With which student do you agree? Why?

Part 2: The Rock Cycle

6) The diagram below shows the three types of rocks. The arrow showing how sedimentary rocks become metamorphic rocks through increasing heat and pressure is drawn below. Draw the rest of the arrows and label them showing how one type of rock can change into another type of rock. Make sure you have an arrow for each of the changes you thought could happen in Questions 2–5. This is a diagram of the rock cycle.

Some of the following terms might be useful: melt into magma, cool from magma into rock, erode into sediments, deposit sediments, transform sediments into rock, increase heat, increase pressure, change.

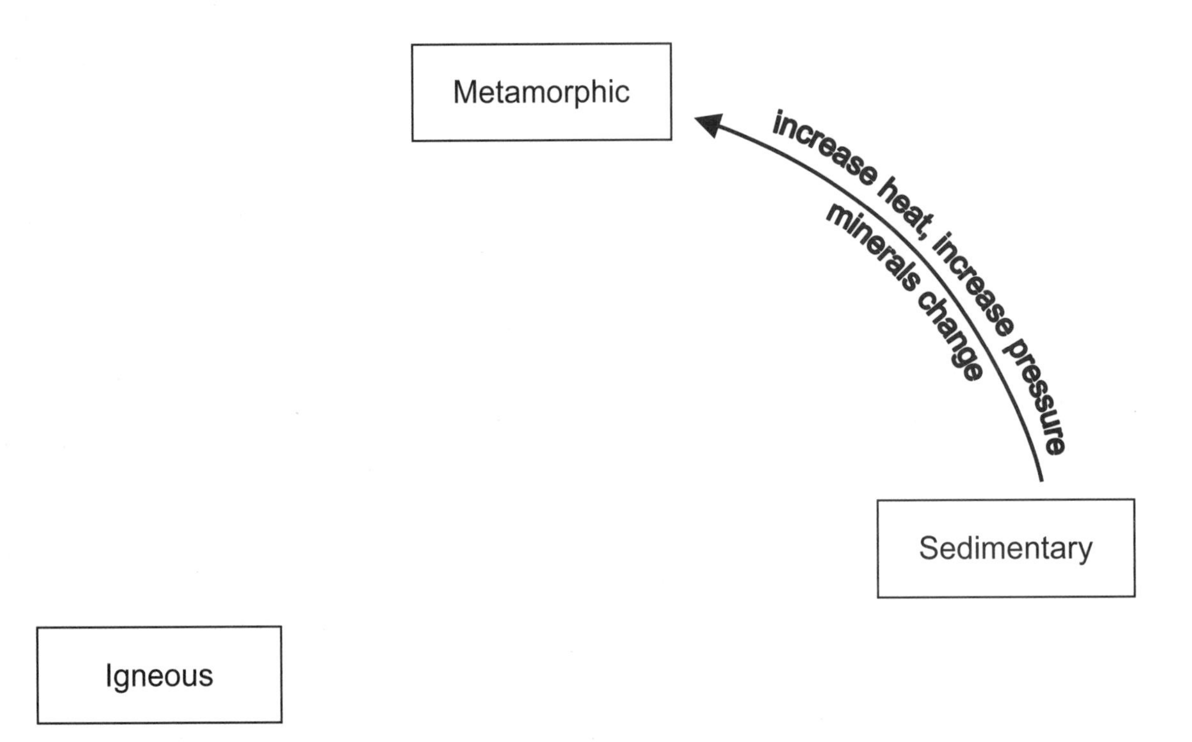

7) Many people describe the rock cycle as rocks going from igneous to sedimentary to metamorphic and then back to igneous again. Explain why this does <u>not</u> fully explain the rock cycle.

Rock Types on Other Planets

On Earth there are igneous, sedimentary, and metamorphic rocks. Understanding the way they formed allows us to determine whether we are likely to find them on other planets.

1) Match the rock type with the correct statement describing its formation.

_____________________ Found where the atmosphere or liquid water causes erosion and movement of rock pieces.

_____________________ Found mostly near convergent tectonic plate boundaries where the temperature and pressure can be very high.

_____________________ Found in places where the interior is so hot that rock melts and then cools again to form new rock.

On the Moon, the first rocks to form when it was molten were the outermost rocks that cooled to form the light-colored highlands. Then molten rock filled in the lower areas and cooled. This rock is called the mare basalt.

2) Based on these descriptions, of what type of rock are the highlands and mare basalt composed?

Highlands: (circle one) igneous sedimentary metamorphic

Mare basalt: (circle one) igneous sedimentary metamorphic

Planet	Water	Atmosphere	Molten interior	Plate tectonics
Mercury	no	no	early only	no
Venus	no	thick	yes	no
Earth	liquid, ice	medium	yes	yes
Moon	no	no	early only	no
Mars	ice	thin	for a while, no more	no

3) Where in our solar system might we find igneous rocks? Explain your choice based on what factors are necessary for an igneous rock to form.

4) Where in our solar system might we find sedimentary rocks? Explain your choice based on what factors are necessary for a sedimentary rock to form.

5) Where in our solar system might we find metamorphic rocks? Explain your choice based on what factors are necessary for a metamorphic rock to form.

6) What is the most common rock type in the solar system? _____________________

Granite Gabbro

Rhyolite Basalt

Jessica Smay

Part 1: Similar Rocks

1) Divide the four igneous rocks shown above into two groups of your choosing. Circle the rocks that are grouped together with this method.

2) What characteristic did you use to determine which rocks belongs in each group?

3) Using different rock characteristics, divide the rocks up again into two different groups.

4) This time, what characteristic did you use to determine which rocks belong in each group?

5) Compare the characteristics you used with other students' characteristics. If you have different characteristics, convince the other students that the two characteristics that you used to divide the rocks are the best two characteristics.

6) After your discussions, list below two ways to divide these rocks into groups.

Identifying Igneous Rocks

Part 2: Color and Mineral Size

The two characteristics that are generally used by geologists to identify igneous rocks are color and mineral size.

7) Fill in the chart below with the appropriate rock names from Part 1.

		Color	
		Light	Dark
Mineral Size	Large		
	Small		

8) The four ovals in the diagram below represent the four different types of igneous rocks shown in the previous questions. Where the ovals overlap, fill in the characteristics shared by both of those rock types.

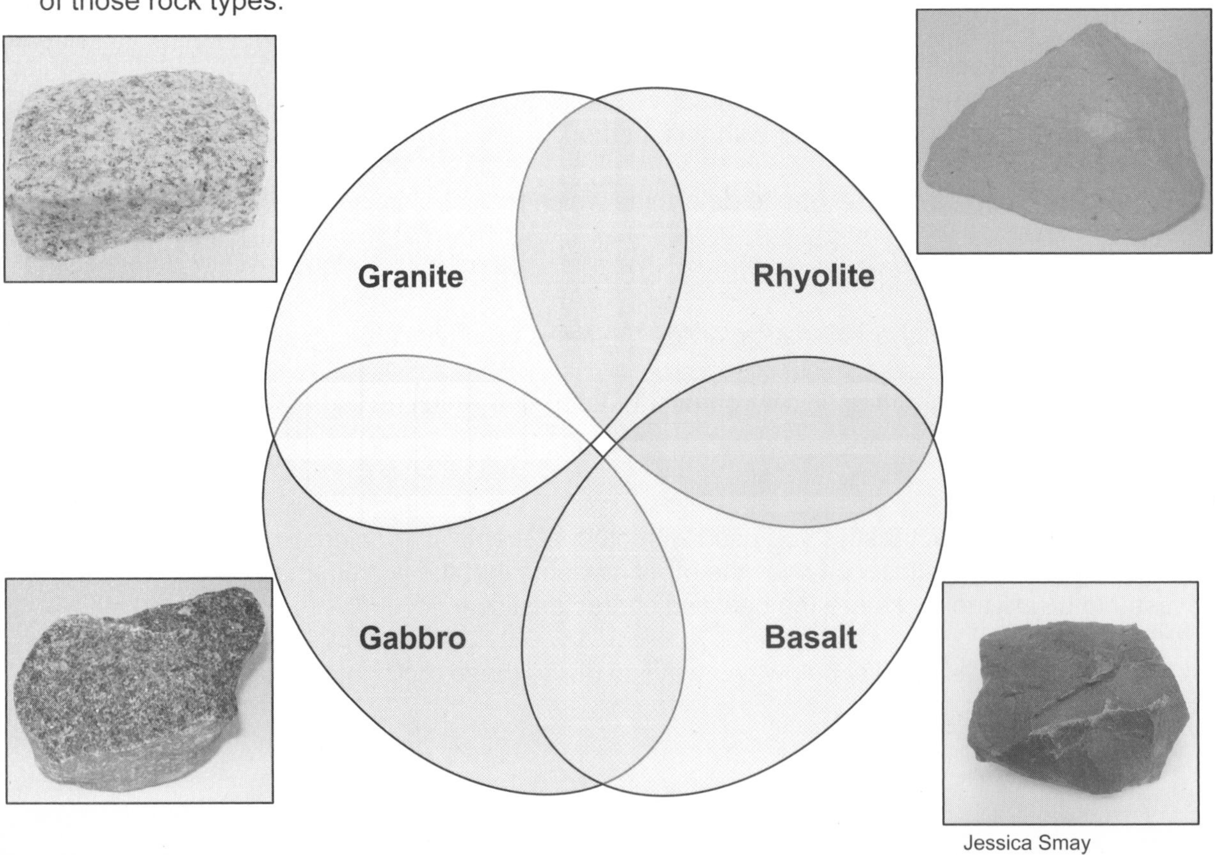

Jessica Smay

Part 1: Forming Minerals
The size of minerals in an igneous rock is determined by how long the magma takes to cool. To illustrate, everyone should stand up and scatter throughout the room.

1) You have two seconds to form groups as big as possible. How many per group? _______

2) Scatter again. Now you have 10 seconds. How many per group? _______

3) Two students are debating about how this activity relates to mineral size in rocks.

Student 1: *It seems to me that with a longer amount of time, it is possible for all the atoms to form really large minerals.*

Student 2: *I don't know, I would think that more time means that more minerals will form, and only a little bit of time means only a few big minerals will form.*

With which student do you agree? Why?

Part 2: Mineral Formation Location
Two bodies of magma are shown in cross section below. One is above ground and the other is deep within the crust. The length of the arrows represents the rate at which heat is escaping from the molten rock as it cools.

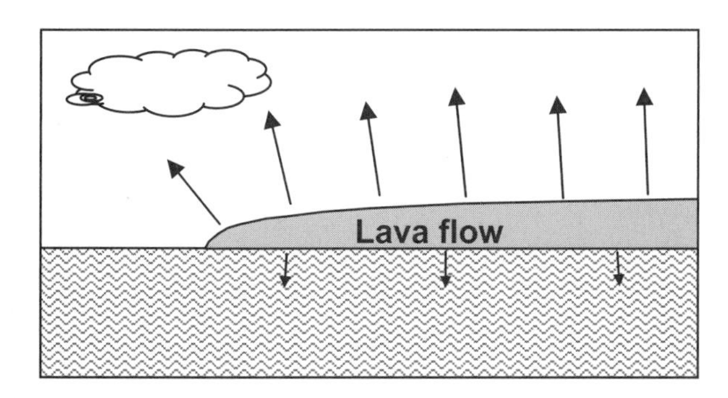

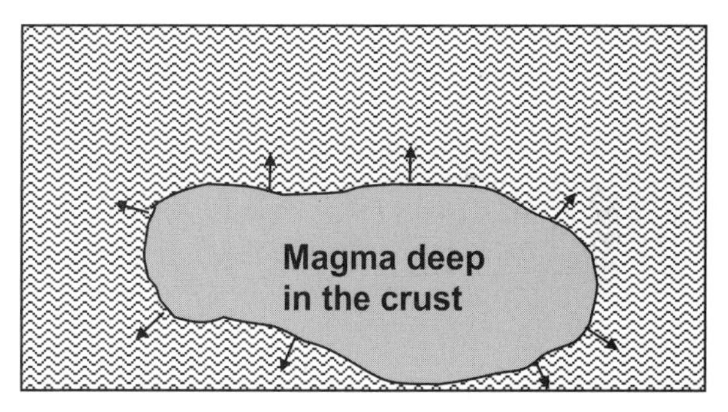

4) Which will cool faster? Lava erupted onto the surface Magma deep underground

5) The igneous rocks granite and gabbro have large minerals. In which location would they have formed?

on the surface deep in the crust

6) The igneous rocks rhyolite and basalt have minerals so small it is difficult to distinguish them with the naked eye. In which location would they have formed?

on the surface deep in the crust

Igneous Rock Mineral Size

7) Circle the two rocks that formed deep in the crust.

Granite

Rhyolite

Basalt

Gabbro

Jessica Smay

Check your answer with your answers for Questions 5 and 6.

Part 3: Porphyry

8) The igneous rock to the right has large, black-and-white colored minerals and many small, gray minerals. You can tell it is an igneous rock because the minerals are rectangular and not rounded like sediments. How might the igneous rock shown to the right have formed?

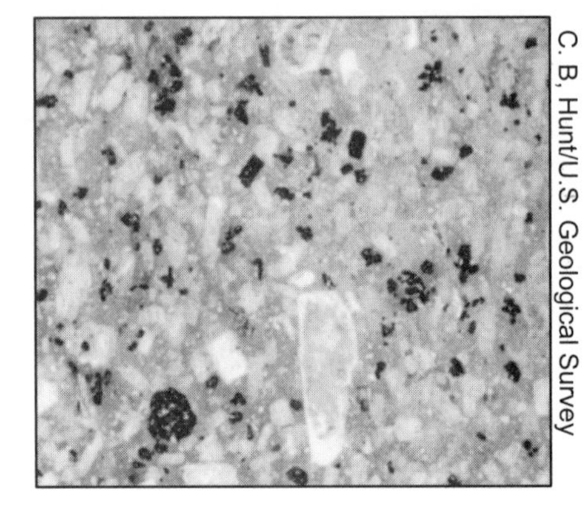

C. B. Hunt/U.S. Geological Survey

Porphyry

9) Two students are debating about the cooling rate of this rock and the formation of the large minerals.

Student 1: *The magma must have gotten large pieces of sediments that we can see trapped in it, and the sediments didn't melt, even though they were in the magma. So, this rock formed because large pieces of sediment got picked up by lava, and then that lava cooled quickly.*

Student 2: *This is an igneous rock, so everything started off as magma. The large minerals must have formed deep underground when the magma was cooling slowly, like in a magma chamber. But the rest of the rock has very small minerals, so they cooled quickly at the surface.*

With which of these students do you agree? Why?

10) Student 2 said that the large minerals formed deep underground, like in a magma chamber, and the small minerals formed at the surface. Describe what actually happened to form the rock. In other words, what story does the appearance of this rock tell us about its history?

(*Hint for Question 10: In what situation is magma in a magma chamber moved to the surface?*)

Different types of lava erupt to form different types of volcanoes. The type of volcano can indicate if the eruptions will be explosive or peaceful.

Felsic lava has a lot of silicon and oxygen chains, and very little iron and magnesium. (Imagine tangled spaghetti strands.)

Mafic lava has a lot of iron and magnesium that break up the silicon and oxygen chains. (Imagine elbow macaroni.)

1) Which type of lava would be runny? mafic felsic

2) Which type of lava would be viscous (thick)? mafic felsic

Below are two simplified profiles of volcanoes. The profile of a volcano can give us a clue if it is built from thick, viscous lava or runny lava (think about toothpaste vs. honey).

3) On the diagram below, write "runny" next to the volcano that is formed from runny lava and "viscous" next to the volcano that is formed from viscous lava.

4) On the diagram below, write "mafic" next to the volcano that is formed from mafic lava, and "felsic" next to the volcano that is formed from felsic lava.

The igneous rock rhyolite is formed from felsic lava, which is why it is light in color. The igneous rock basalt is formed from mafic lava, which is why it is dark in color.

5) On the diagram below, write "rhyolite" next to the volcano composed of rhyolite, and "basalt" next to the volcano composed of basalt.

Volcanoes will erupt explosively if the gasses in the lava cannot easily escape. Runnier lava will allow the gas to bubble out peacefully, so there is no explosion. However, viscous lava will not allow the gas to escape, so the pressure builds until an explosion releases the gas.

6) Which type of lava will easily let gasses escape? felsic lava mafic lava

7) Write "explosive" next to the volcano that would erupt explosively and "peaceful" next to the volcano that will erupt peacefully.

Shield volcano

Composite volcano

Not to scale

8) Based on the type of eruption, which volcano would you rather live next to—a shield volcano or composite volcano? Explain your answer.

9) Two students are debating which volcano they would rather live next to.

Student 1: *I would prefer to live next to a composite volcano because the lava is so thick and viscous that they don't have long lava flows. The lava flows will slow down and stop before they get to my house.*

Student 2: *But, composite volcanoes have explosions that can't be predicted, so they are more dangerous than runny lava. I would prefer to live next to a shield volcano, where the runny lava flows in predictable patterns.*

With which of these students do you agree? Why?

10) The photograph below is of a volcano in the United States. What can you determine about the volcano based on the picture (e.g., type of volcano, lava, rock, eruption...)?

Hoblitt/U.S. Geological Survey

Part 1: Observing Volcano Distribution
When a scientist makes a discovery, it helps to have as many different sources of information as possible confirm that discovery. Here we will look at two ways to determine the types of volcanoes on other planets.

1) Examine the maps of volcanoes on Mars, Venus, and Earth. Take one minute to determine if there is a clear pattern in the location of volcanoes on each planet, or if they are distributed in random groups. If there is a pattern, describe what kind of pattern you see.

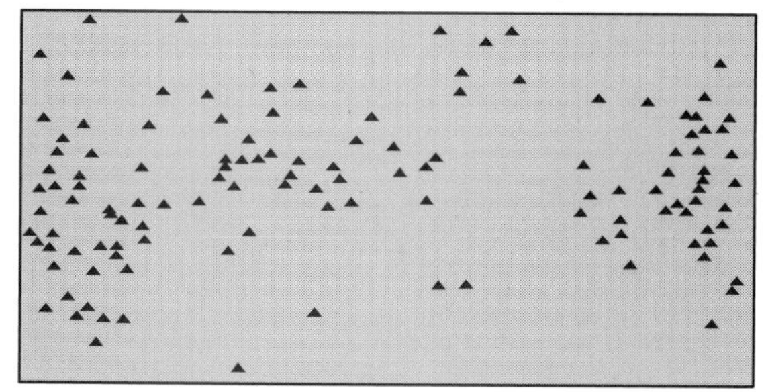

Volcanoes on Venus (triangles)

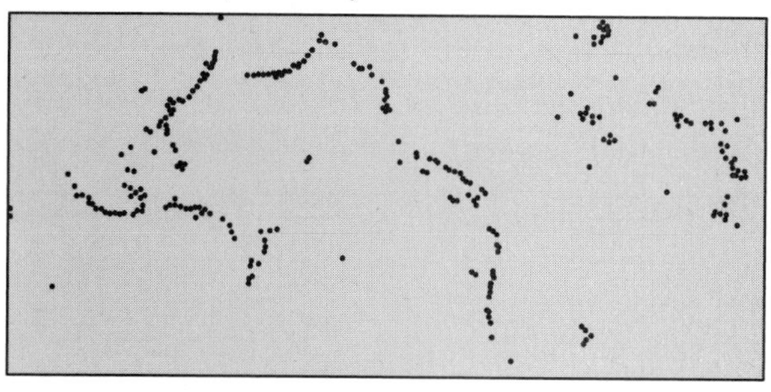

Volcanoes on Earth (dots)

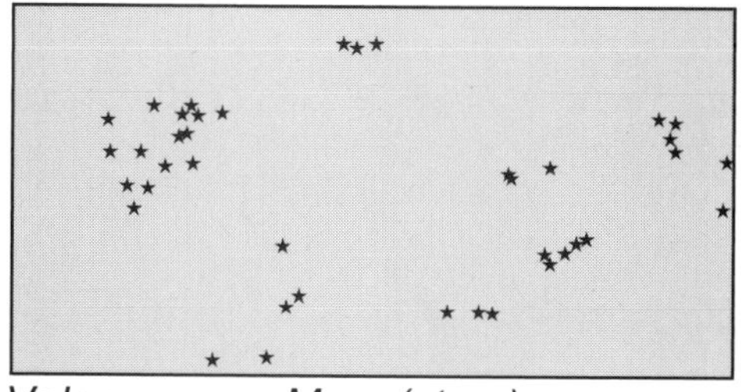

Volcanoes on Mars (stars)

Part 2: Analysis
Volcanoes form either at randomly distributed hotspots or lined up along tectonic plate boundaries. A single planet might have both types of volcanoes.

2) Why do the volcanoes on Earth form where they do? hot spots plate tectonics

Explain how your answer is related to your observations about the maps.

3) Why do the volcanoes on Venus form where they do? hot spots plate tectonics

Explain how your answer is related to your observations about the maps.

4) Why did the volcanoes on Mars form where they did? hot spots plate tectonics

Explain how your answer is related to your observations about the maps.

5) Which planet(s) has/have plate tectonics? Venus Earth Mars

Volcanoes on Other Planets

Part 3: Comparing Individual Volcanoes

Another way to determine the cause of volcanoes on other planets is to compare the two types of volcanoes on Earth with volcanoes on other planets. Composite volcanoes (e.g., Mount St. Helens) usually form at plate tectonic boundaries and have steep slopes; shield volcanoes (e.g., Hawaii) usually form at hot spots and have gentle slopes.

6) Look at the profile of volcanoes on Earth drawn to scale. Label each volcano as “composite volcano” or “shield volcano” and indicate if the volcano formed at a hot spot or plate tectonic boundary.

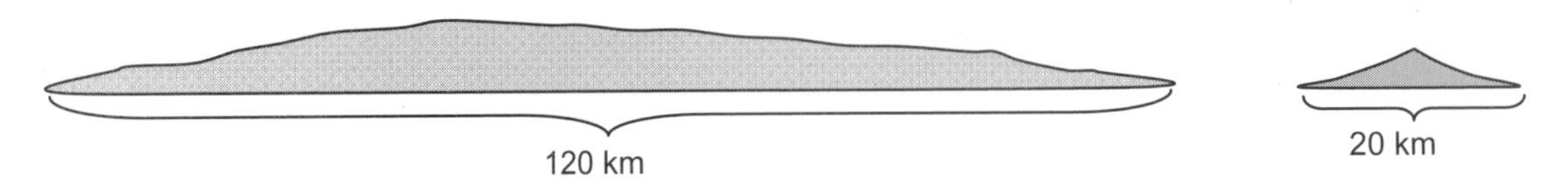

Below is a satellite image of Olympus Mons, an example of a volcano on Mars. This volcano is approximately 25 km tall and 600 km wide. It is possible to use satellite information to create a profile like those of Earth shown above.

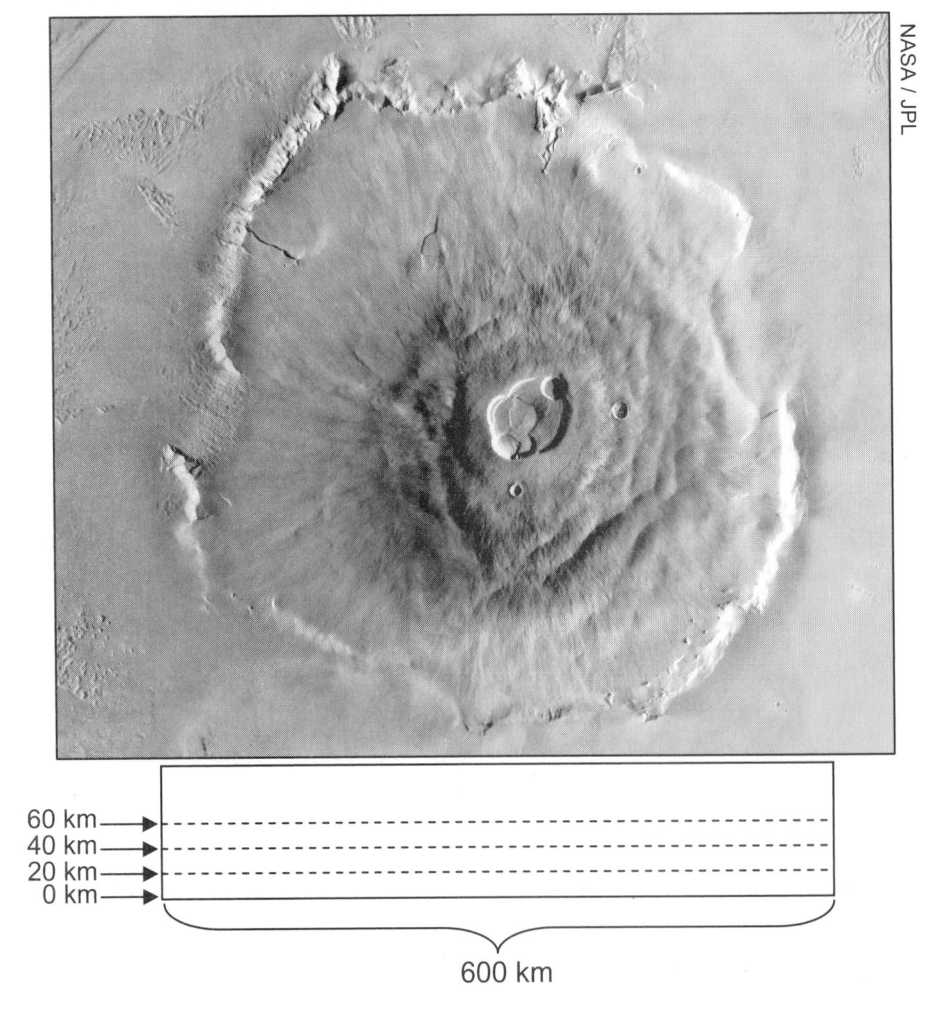

7) Use the information about the height and width of the volcano to sketch the profile of this volcano (like those of volcanoes on Earth shown in Question 6) on the graph above.

8) Based on the profiles, is the volcano on Mars a composite volcano or a shield volcano? Explain.

9) Based on the profiles, why did the volcanoes on Mars form? hot spots plate tectonics

Explain how your answer is related to the profile and type of volcano.

You used two methods to determine the type and origin of volcanoes found on Mars: the distribution of volcanoes to determine the likely source of volcanism, and the profile of an individual volcano.

10) Do your two data sets agree? If they do not agree, what might cause the difference?

11) Why is it helpful for a scientist to have two or more different data sets when giving evidence to support a discovery?

Part 1: Creating Sedimentary Rocks

1) From the choices, list the steps necessary for a parent rock to become a detrital or chemical sedimentary rock. Each step will be used once.

DETRITAL (e.g., shale/mudstone)

- transportation (water, wind, ice)
- ~~parent rock is broken into smaller pieces~~
- deposition of detrital sediments
- compaction and cementation into rock

CHEMICAL (e.g., limestone)

- transportation (dissolved in water)
- ~~parent rock is dissolved~~
- precipitation as rock
- precipitation as shells
- shells are deposited, compacted, and cemented

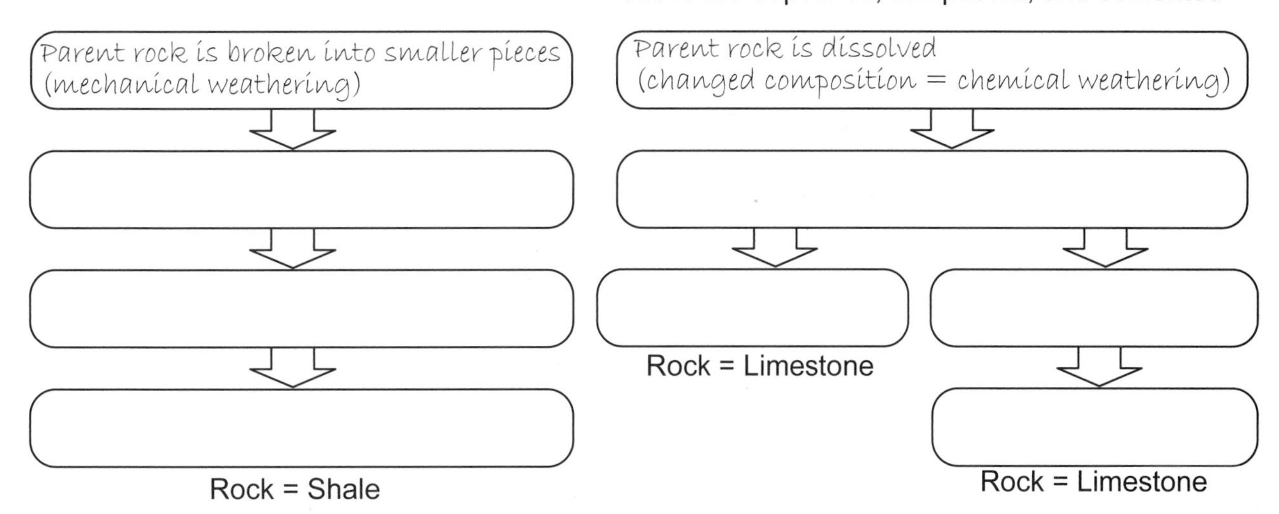

Part 2: Changing from Sediments to Sedimentary Rocks

2) Sandstone is a detrital sedimentary rock made up of sand-sized pieces. Examine the four steps and draw arrows to where they are occurring in the diagram.

- parent rock is broken into sand-sized pieces (sediments)
- sand-sized sediments are transported
- sand-sized sediments are deposited and buried
- deposited and buried sand is compacted and cemented into sandstone (no longer sediments)

Part 3: Sediments vs. Sedimentary Rocks

3) One of these photos is a photo of pebble-sized sediments, the other is of a single sedimentary rock. Label the photos. There is a finger for scale in the left photo.

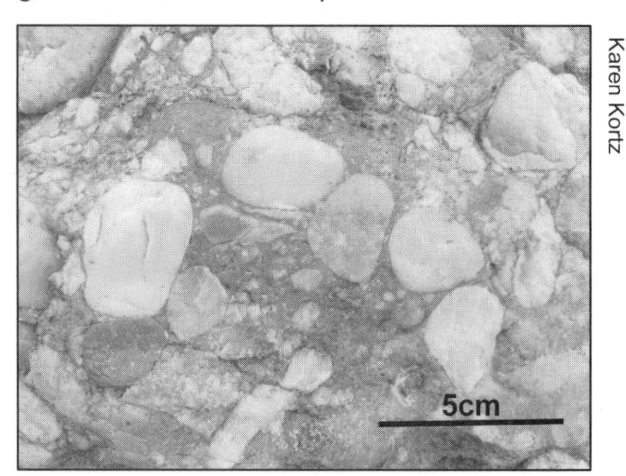

4) What needs to happen so that pebble-sized sediments become a sedimentary rock?

5) Two students are debating about the pebbles in the image on the left and what type of rock the pebbles are.

Student 1: *In the photo on the left, it is clear that the rocks there are all pebble sized. I think that each pebble is a sedimentary rock because the pebbles have been broken from much larger rocks and they were transported to this new area.*

Student 2: *I don't agree. Each pebble can be any type of rock—igneous, metamorphic, or sedimentary. Because they have not yet been compacted and cemented together, they have not yet been turned into a sedimentary rock like in the photo on the right.*

With which student do you agree? Why?

6) You are at the beach with a friend and find pebbles and some slightly larger rocks. Explain why you would not call these sedimentary rocks.

Part 1: Rock Types and Features in Different Environments

1) Match the picture of the sedimentary rock to the name of the rock and to the best answer for the environment of deposition. Do this by connecting them with a line.

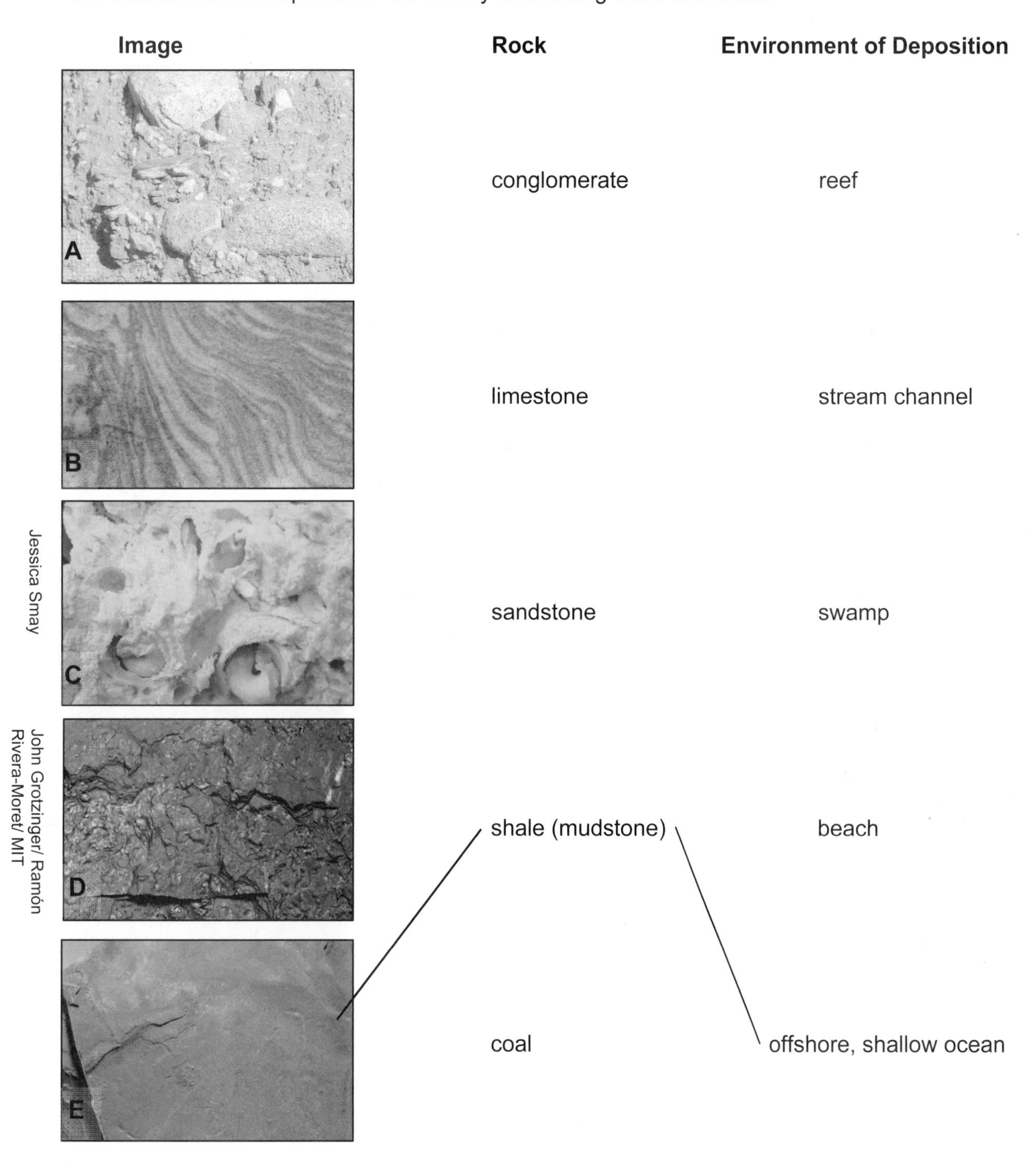

2) Coal mines are common in Pennsylvania. What was the environment of Pennsylvania in the past compared to today?

Circle the environment that best matches what Pennsylvania was like in the past.

Karen Kortz
Karen Kortz
C. W. Cross/ U.S. Geological Survey
Karen Kortz

3) There is a thick layer of limestone with many shells under Chicago. What can you figure out about the geologic past of that area?

Depositional environments vary from location to location. A particular place on Earth's surface can also change depositional environments over time.

1) An area has a layer of sandstone that is interpreted to have once been a beach. On top of that sandstone is a layer of limestone that formed deep in the ocean. What can you say about the sea level of this area?

Limestone
Sandstone

2) Two students are debating about how layers of sandstone and shale (mudstone) relate to rising sea level.

Student 1: *Sandstone forms right at the ocean's edge and shale forms in deeper water. So if the water was shallow first, and then it got deeper, sandstone would be deposited first on the bottom, and shale would be on top.*

Student 2: *But if the water is getting deeper, that means the shale should be deeper because it forms in deeper water. So, I think the shale would be underneath the sandstone because it forms in deeper water.*

With which student do you agree? Why?

3) The diagram below shows three rock layers. Which rock layer formed first?

bottom middle top

4) If a rock layer formed on the land, what are possible rocks that make up the layer?

conglomerate limestone sandstone shale (mudstone) coal

5) If a rock layer formed in deep water, what are possible rocks that make up the layer?

conglomerate limestone sandstone shale (mudstone) coal

6) In the diagram below, write a series of three rock types that could form in an area where the sea level is rising over time. The area started on land and was covered by deeper and deeper water over time. Explain your sequence.

Top rock layer:	
Middle rock layer:	
Bottom rock layer:	

Part 1: Metamorphic Changes

When rocks are compressed and heated within Earth, they change, or metamorphose. During metamorphism, rocks tend to change in the following ways:

- minerals change into different, more stable minerals
- minerals grow larger
- flat minerals align themselves parallel to each other

1) Examine the photos below. Describe one or two observations how the minerals in these rocks changed during metamorphism.

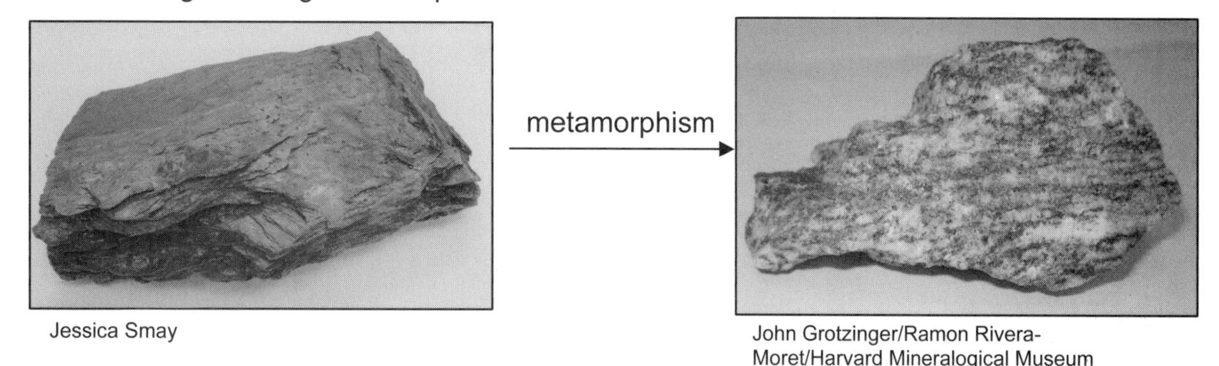

Jessica Smay

John Grotzinger/Ramon Rivera-Moret/Harvard Mineralogical Museum

Part 2: Metamorphic Grades

During metamorphism, the temperature of and pressure on a rock determine what the rock looks like.

If the temperatures and pressures are not very different from those on the surface of Earth (low-grade metamorphism), the minerals do not change much and are small.

In high temperatures and pressures (high-grade metamorphism), the minerals in a metamorphic rock change a lot, so they grow large and tend to form into dark and light bands.

Slate Jessica Smay

Schist

Gneiss

2) List the rocks above in order from low-grade to high-grade metamorphic rocks.

Low grade ________________ ________________ ________________ **High grade**

3) Describe what slate would look like if it was metamorphosed more.

4) Three students are thinking about the changes in slate during metamorphism.

Student 1: *I think it would change to a higher metamorphic grade, so instead of being a low-grade metamorphic rock, it would change to an intermediate-grade metamorphic rock. So it will become schist, and it would have larger minerals.*

Student 2: *It would look pretty much the same, but it would get even flatter because of the higher pressures during metamorphism. For example, if it started out two inches thick, it might be one inch thick after metamorphism.*

Student 3: *It would melt, and then new minerals would form.*

With which student do you agree? Why?

Part 3: Forming Metamorphic Minerals

5) Two students are thinking about how metamorphic minerals form.

Student 1: *In metamorphic rocks, the minerals melt a little bit due to the extremely high temperatures. The melting causes the atoms to flow around and grow bigger minerals.*

Student 2: *I thought that new minerals form and grow bigger because the rock is getting compressed and heated. The atoms that formed minerals in the parent rock rearrange to form bigger, new minerals.*

With which student do you agree? Why?

6) Schist contains a lot of the mineral mica, which is what makes it shiny. Slate contains clay, but not much mica. Where do the mica minerals come from as slate metamorphoses into schist?

7) If a rock melts, can it be considered a metamorphic rock? Explain.

8) Use your answer to Question 7 to change your answers to Questions 4 and 5, if necessary.

The History of Metamorphic Rocks

Part 1: Parent Rocks

The parent rock of a metamorphic rock is the original rock before it was changed by metamorphism. It plays an important role in determining the type of resulting metamorphic rock.

1) For each pair below, fill in the parent rock or the resulting metamorphic rock as needed. The rocks we will be considering are limestone, shale (mudstone), marble, slate, and gneiss.

2) A large area undergoes the same amount of metamorphism (all rocks reach the same metamorphic grade). However, after the metamorphism, some rocks are marble and some rocks are slate. Why?

3) There is a large area made up of schist. What is the most likely environment that existed before the area was metamorphosed?

desert dunes   coral reef   ocean floor   volcano

4) A metamorphic rock is a "changed" rock. How is it possible to use metamorphic rocks to figure out the geologic history of an area before the area was metamorphosed? Give at least one example in your answer.

The History of Metamorphic Rocks

Part 2: Metamorphism and Plate Boundaries

Two common types of metamorphism are:

Contact metamorphism: high temperatures from being near hot magma

Regional metamorphism: high temperatures and pressures associated with mountain building

For each of the tectonic locations below, circle the type(s) of metamorphism that might occur. Use the space to write a brief explanation.

5) divergent boundary: contact regional none

6) convergent boundary (ocean-continent): contact regional none

7) convergent boundary (continent-continent): contact regional none

8) transform boundary: contact regional none

9) The metamorphic rock gneiss forms primarily through regional metamorphism. At what type of ancient plate boundary(ies) would you expect to find gneiss?

10) Two students are discussing what they could determine about a region that is composed of a large amount of marble.

Student 1: *We could say that the area was once a shallow sea because marble forms from limestone, and limestone forms in shallow seas.*

Student 2: *We could say that the area was once a convergent plate boundary because the rocks are metamorphosed. If there was no convergent boundary, there is no way that all the rocks would have been metamorphosed.*

Do you agree with one or both students? Why?

11) In New England, you can find marble, slate, and gneiss. What can you interpret about the geologic history of the area?

Below is a topographic map of a small island. Use this topographic map to answer the questions. The outer topographic line is at sea level.

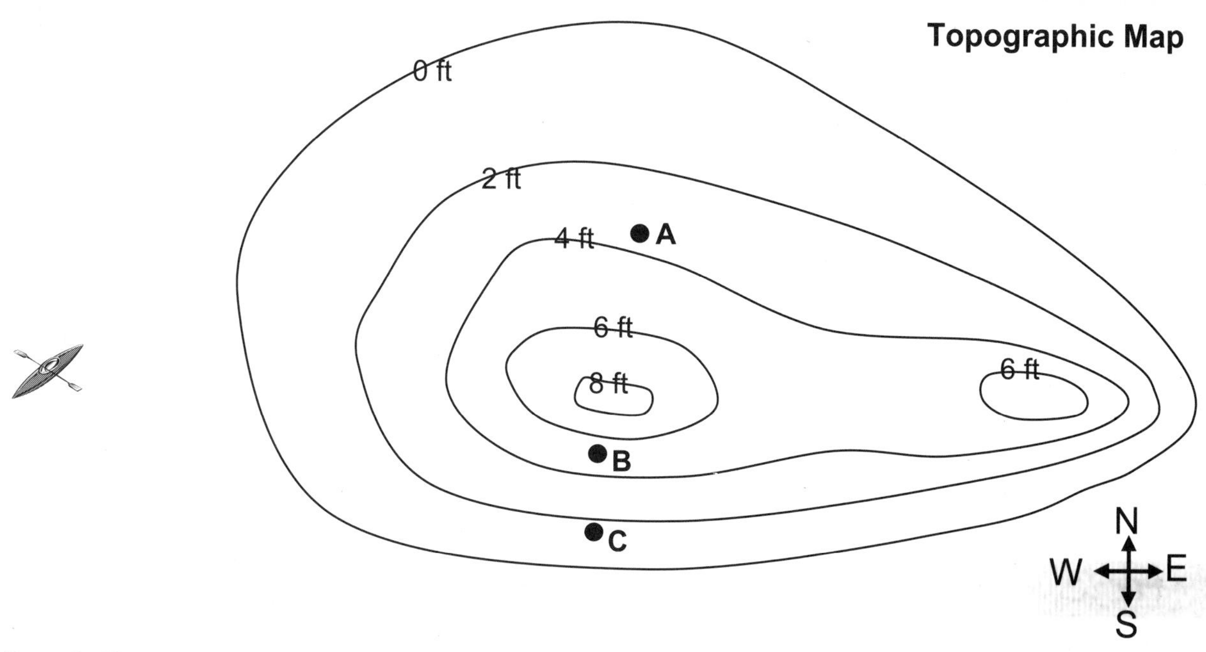

Part 1: Topographic Maps and Steepness

1) Which side of the island is steeper, the west side or the east side? Explain your answer.

2) Two students are debating which side of the island is steeper.

Student 1: *I think the west side is steeper because it goes from the highest elevation down to sea level. On the east side, the elevation change is smaller because that hill is only 6 feet tall. The big change in elevation on the west side means it's steeper.*

Student 2: *No, I disagree. It is the east side that is steeper because the steepness is shown by how close contours are together. If they're closer, that means the elevation is changing faster, which means it's steeper.*

With which student do you agree? Why?

Part 2: Profiles

3) You walk straight from B to C. Would you walk uphill, downhill, or a combination of both?

4) You walk straight from B to A. Would you walk uphill, downhill, or a combination of both?

5) You are in a canoe far to the <u>west</u> of the island (see map). You look back to the east at the profile of the island. Which sketch below best shows what the island would look like? (Circle it.) Explain why the other two choices are not correct.

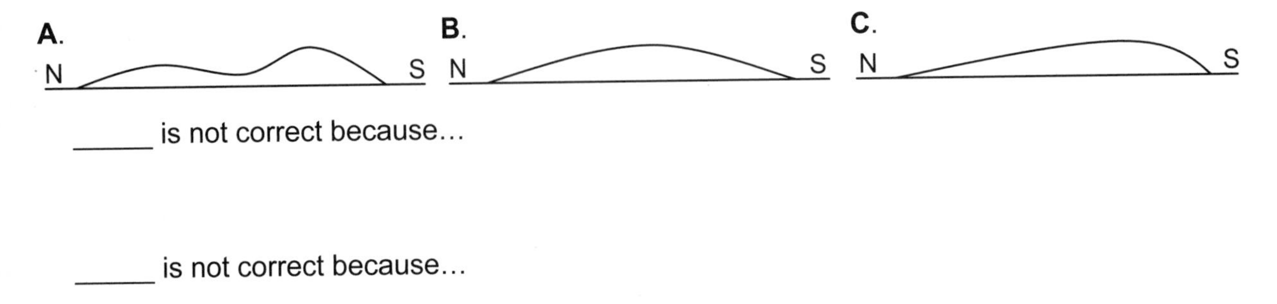

_____ is not correct because...

_____ is not correct because...

Below is an east-west profile across the island (from the south looking to the north). Use this profile to answer the following questions.

Profile

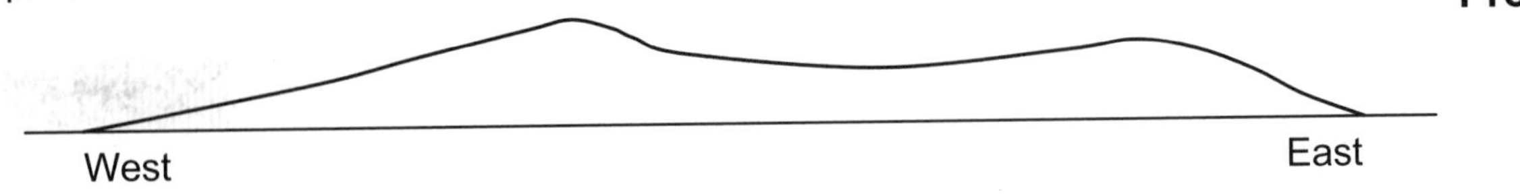

6) Is it possible to show an arrow indicating "up" on the profile? If it is possible, then draw in the arrow. If it is not possible, explain why not.

7) Is it possible to show an arrow indicating "north" on the profile? If it is possible, then draw in the arrow. If it is not possible, explain why not.

8) Two students are debating what arrows can be drawn on profiles.

Student 1: *I think that both the arrows can be drawn on the profile because "up" and "north" on maps are similar. You could draw them both pointing in the same direction toward the top of the paper.*

Student 2: *"Up" is the direction away from ground, and it would point toward the top of the paper, like you said. But "north" and "up" are not the same direction because "north" on the profile points into the paper, so it cannot be drawn.*

With which student do you agree? Why?

Part 1: Planet Features

1) What features can form on a planet if it has a hot, molten interior?

dunes   impact craters   stream beds   volcanic lava flows

2) What features can form on a planet if it has an atmosphere?

dunes   impact craters   stream beds   volcanic lava flows

3) What features can form on a planet if it has liquid on the surface?

dunes   impact craters   stream beds   volcanic lava flows

4) What features can form on a planet with any characteristics?

dunes   impact craters   stream beds   volcanic lava flows

The following images are of three different Earth-like planets (Mercury, Earth, and Mars). All the images are of the solid, rocky surface and were taken by NASA spacecraft.

5) Examine these images and identify the type of <u>surface feature</u> shown: sand dunes, impact craters, stream beds, lava flows.

6) For each planet, write down what you can determine about the planet based on those images (if it has an atmosphere, a molten interior, or liquid on the surface).

Planet 1

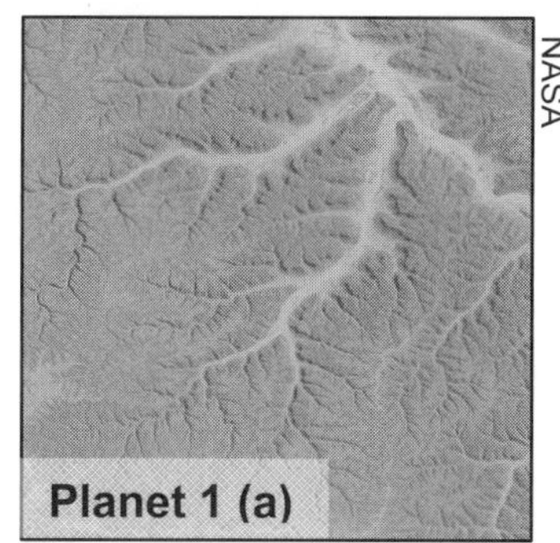

NASA

Feature: Stream beds

Planet 1 (b)

NASA/GSFC/METI/ERSDAC/JAROS, and U.S./Japan ASTER Science Team

Feature:

NASA/JPL/UCSD/JSC

Feature:

What I know about this planet:

Planet Surface Features

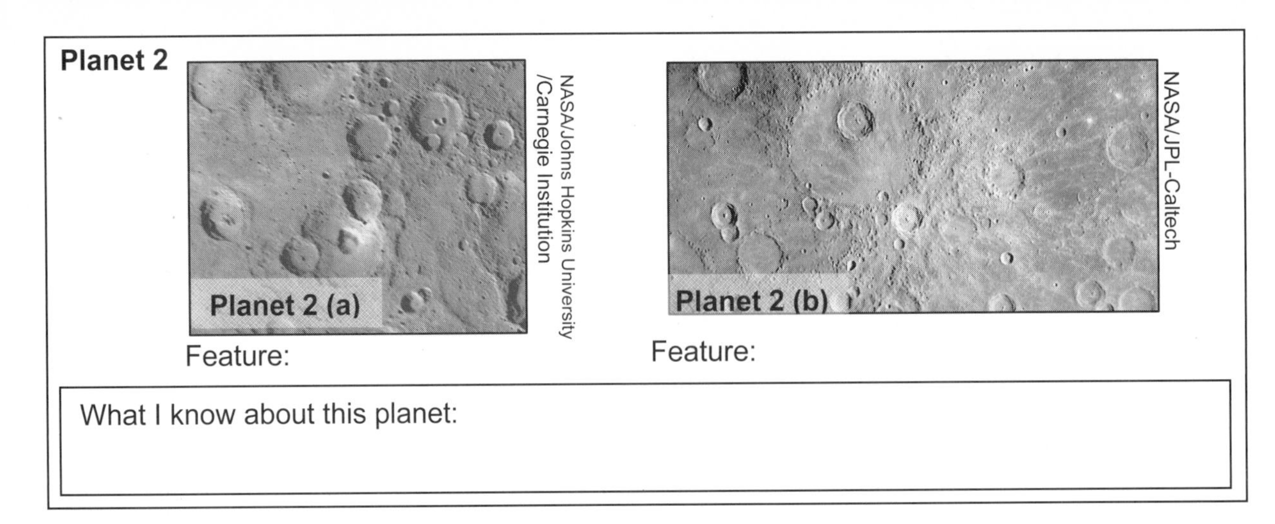

Feature:

Feature:

What I know about this planet:

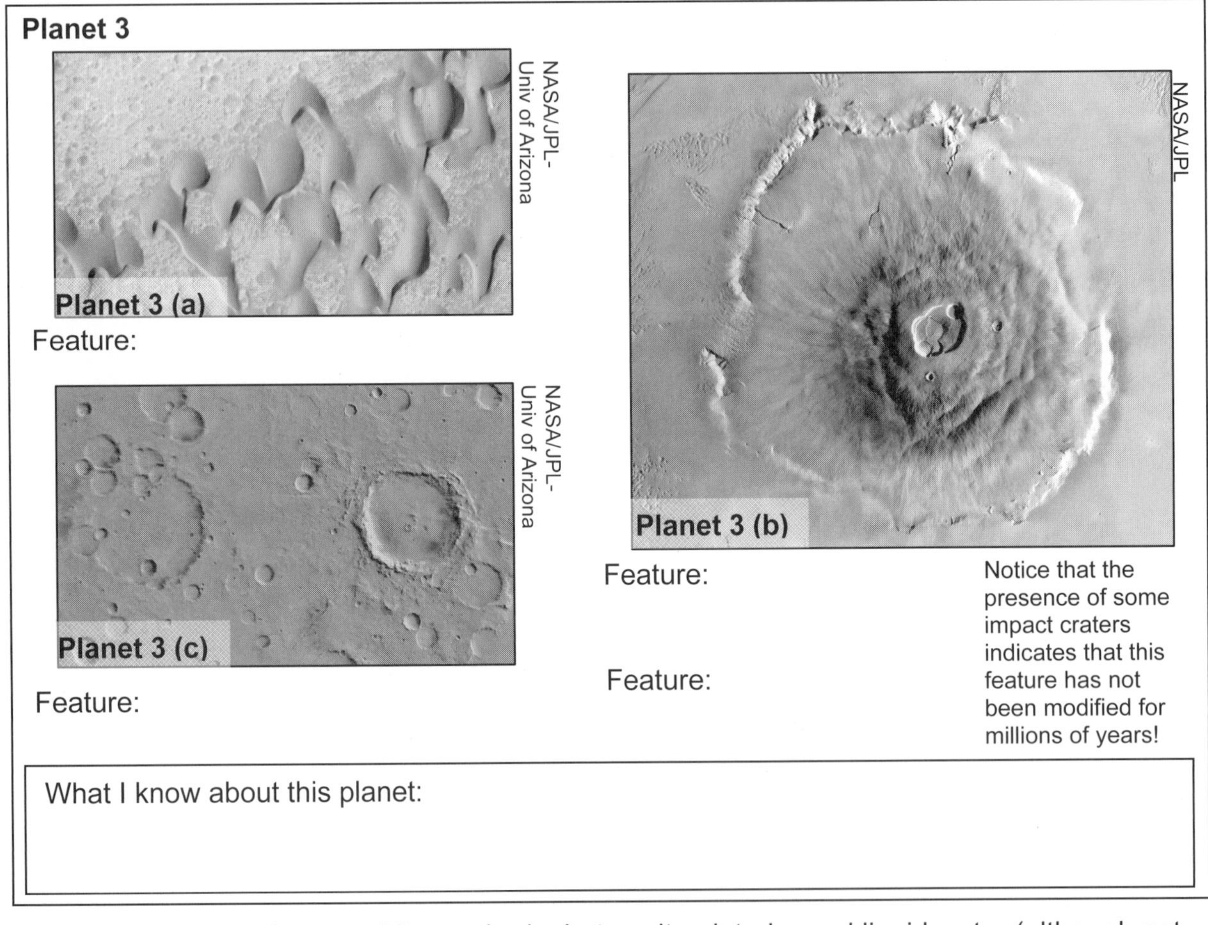

Feature:

Feature:

Feature:

Feature:

Notice that the presence of some impact craters indicates that this feature has not been modified for millions of years!

What I know about this planet:

Mars has an atmosphere, and it once had a hot, molten interior and liquid water (although not any more). Mercury does not have an atmosphere, liquid water, or a hot, molten interior.

7) Using the information given, label each of the planets as Earth, Mercury, or Mars.

8) The Moon is completely covered in craters. What can you determine about the Moon based on this information?

Part 1: Magnitude vs. Intensity

The magnitude of an earthquake is the amount of energy that is released as the rock breaks. It is the Richter scale number generally displayed by the news.

The intensity of an earthquake is the measure of damage and deaths it caused. A high-intensity earthquake results in a great deal of damage and a high death toll.

1) Describe two situations in which a large-magnitude earthquake can have a low intensity.

2) Describe two situations in which a small-magnitude earthquake can have a high intensity.

3) Two students are debating earthquake intensity and magnitude.

Student 1: *An example of a large-magnitude earthquake that has a low intensity is if it hit an area with a low population.*

Student 2: *So, if a really, really large-magnitude earthquake hit the desert in California where nobody lives, you're saying it would not have a high intensity? I don't agree. If the earthquake is that big, I think it would need to have a high intensity as well.*

With which student do you agree? Why?

Part 2: Earthquake Intensities

The Loma Prieta earthquake (1989) caused 63 fatalities and $10 billion in damage. During this large 7.1 magnitude earthquake, there was heavy damage near the fault line where it moved, but there was more damage approximately 50 miles away at the edges of heavily populated San Francisco Bay. The buildings and bridges that were built on soft sediments near the Bay were destroyed or damaged.

H.G. Wilshire / U.S. Geological Survey

4) The Loma Prieta earthquake is an earthquake with a high intensity. Explain why the earthquake had such a high intensity. There will be more than one reason.

Part 1: Earthquake Patterns

The USGS map below shows the locations of earthquakes around the world over a 27-year period.

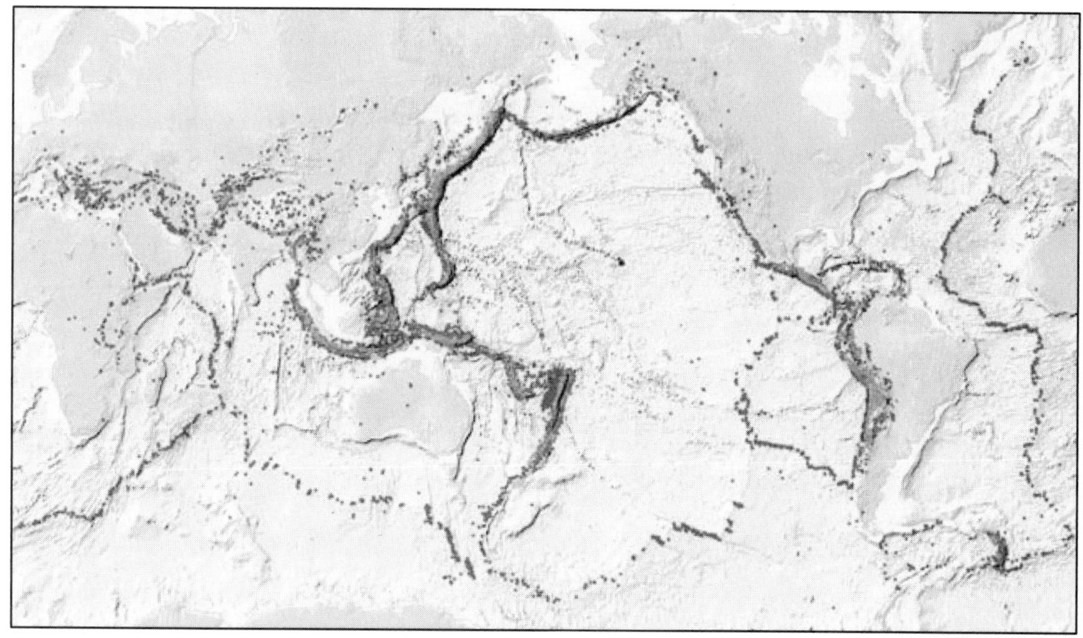

Take a careful look at the pattern of earthquakes scattered around Earth's surface.

1) Do earthquakes occur around the coastlines of all oceans? Yes No

If you chose "No," give an example of an ocean with no earthquakes around the edges.

2) Do earthquakes occur just along coastlines? Yes No

If you chose "No," give an example of an area where earthquakes occur in the middle of an ocean, not on a coastline.

3) Do earthquakes occur just in hot climates? Yes No

If you chose "No," give an example of a cold area that experiences earthquakes.

4) Why do earthquakes occur where they do?

5) Two students are debating about why more earthquakes occur in California than in New York.

Student 1: *Earthquakes occur where the faults are, so areas that have lots of faults also have lots of earthquakes. California has a lot of faults, so it has a lot of earthquakes, unlike New York.*

Student 2: *That only partly answers the question because we need to know why faults occur where they do. Most faults occur along plate boundaries, so it's because California is on a plate boundary and New York is not.*

Do you agree with one or both students? Why?

Part 2: Earthquakes and Plate Boundaries
The map to the right shows the locations of earthquakes in and near South America.

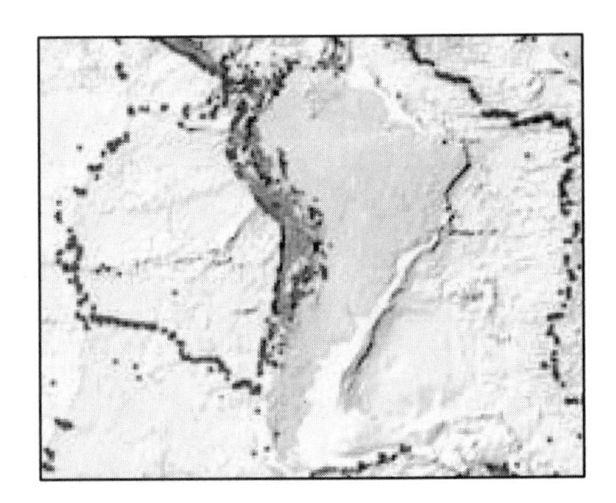

6) Based on the locations of the earthquakes, draw three or four lines on the map indicating where the major plate boundaries are located.

The map below shows the locations of plate boundaries in North America as solid grey lines.

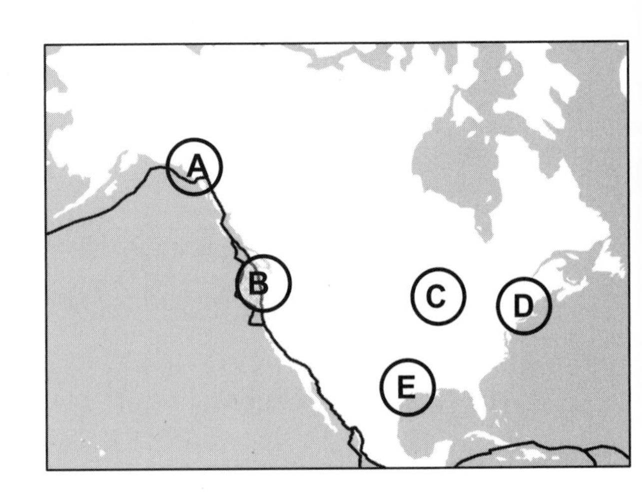

7) Based on the locations of the plate boundaries, what locations are least likely to experience earthquakes?

A B C D E

8) Do all locations on the coast commonly experience earthquakes? Explain your answer using examples.

9) Do warm areas more commonly experience earthquakes than cold areas? Explain your answer using examples.

10) Explain to your friends why some places in North America get more earthquakes than others.

Part 1: Plate Boundaries and Tsunami

Most tsunami form when an earthquake occurs underwater. To create a tsunami, this earthquake needs to (1) be large, and (2) move the seafloor up or down.

Below are cross-section diagrams showing the movement as a result of earthquakes that occur at each of the plate boundaries. The large, black arrows show the movement of the plate.

1) Label each diagram of a fault with the correct plate boundary at which it normally occurs: divergent, convergent, or transform.

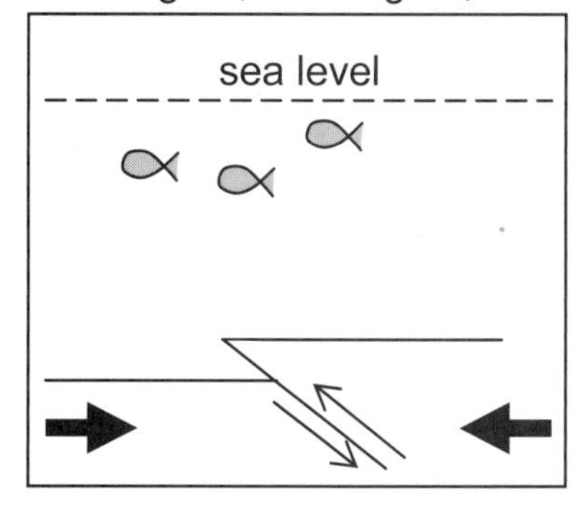

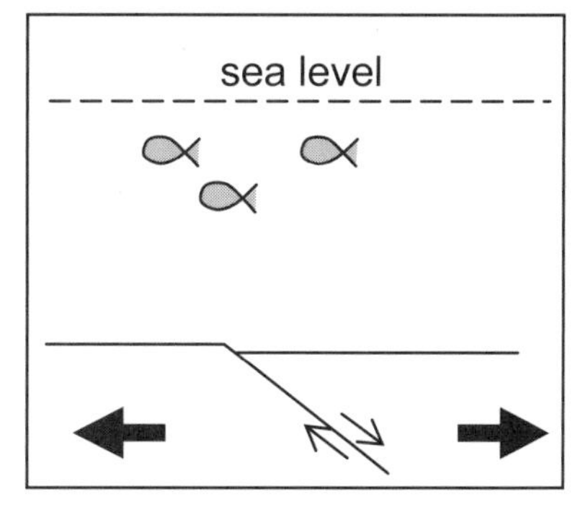

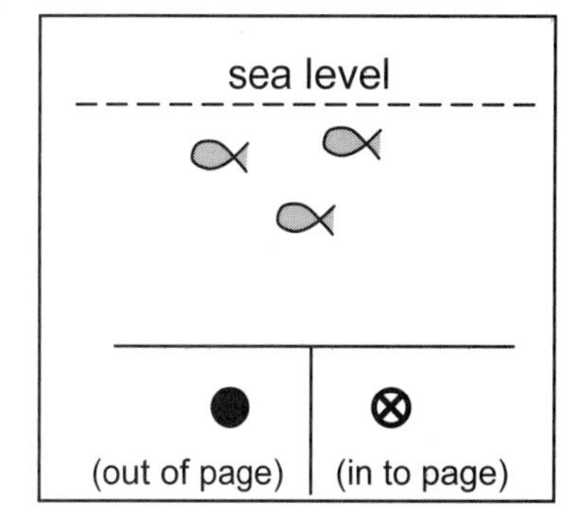

2) The largest earthquakes occur when rocks are compressed. Which type of plate boundary experiences the largest earthquakes?

divergent   convergent   transform

3) At which type(s) of plate boundary do earthquakes cause the seafloor to move up and down?

divergent   convergent   transform

4) Using the two criteria described above, which diagram(s) above illustrates where tsunami usually form?

divergent   convergent   transform

5) Two students are debating about where tsunami usually form.

Student 1: *Tsunami usually only form at convergent plate boundaries because that's the plate boundary where the largest earthquakes happen and the seafloor moves up or down during the earthquake.*

Student 2: *I think tsunami form at any plate boundary when any earthquake occurs underwater, as long as the earthquake is large enough. For example, California gets lots of earthquakes, and it is near the coast, so the San Andreas Fault, which is a transform fault, can cause tsunami.*

With which student do you agree? Why?

Part 2: Tsunami Formation

The sequence of diagrams to the right shows the formation of a tsunami along the edge of a continent.

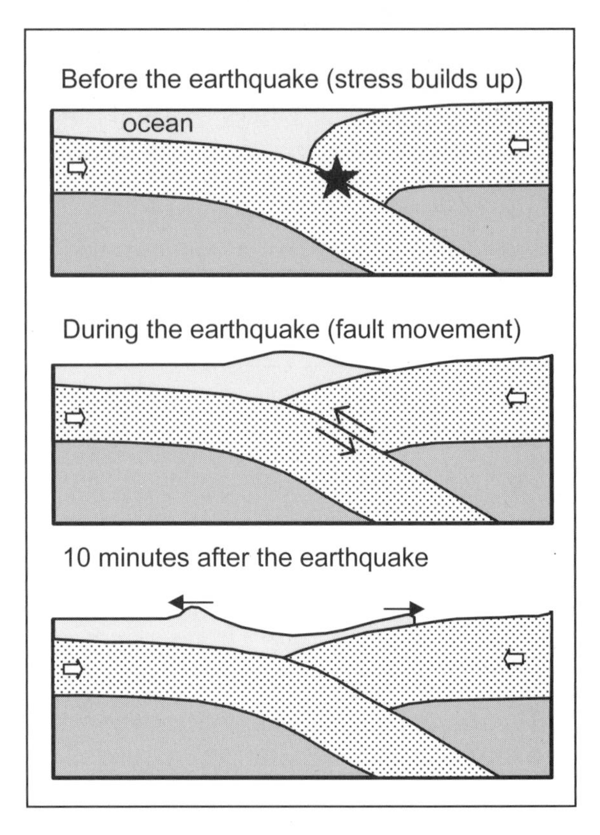

6) What type of plate boundary is shown in the diagrams?

convergent divergent transform

7) Why is there no movement between the plates at the star before the earthquake?

8) What happens to the ocean water when the earthquake occurs?

The map below shows the locations of plate boundaries. Recall that divergent boundaries tend to occur in the middle of oceans, whereas convergent boundaries tend to occur near the edges of oceans.

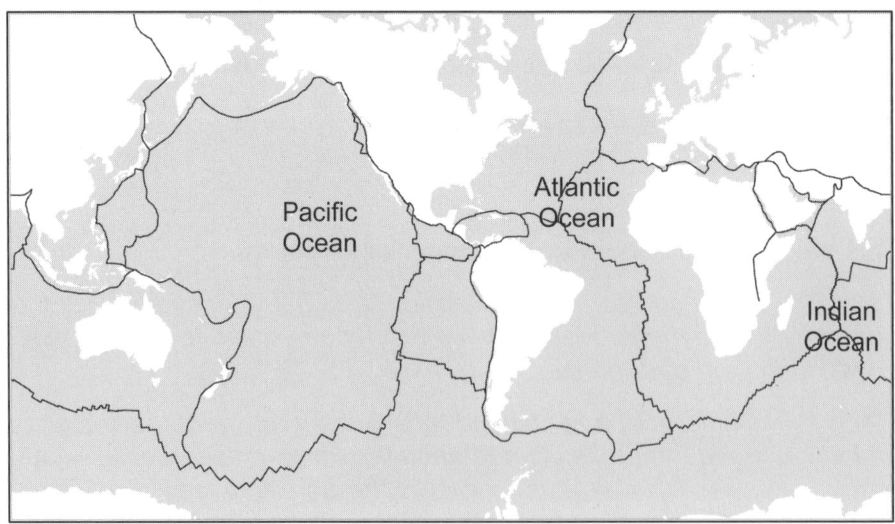

9) Which ocean has the most tsunami? Atlantic Pacific Indian

Explain your answer.

Part 1: Constructing a Curve

A flood-frequency curve plots the discharge of a particular stream against how often that discharge occurs. In the graph below, the levels of floods of a stream in the United States were recorded each year between 1950 and 1999.

Year	Flood * discharge	Year	Flood * discharge	Year	Flood * discharge	Year	Flood * discharge	Year	Flood * discharge
1950	**18**	1960		1970	**33, 18**	1980		1990	**50**
1951	**50**	1961	**18**	1971		1981	**18**	1991	**18**
1952		1962	**18, 33, 18**	1972	**33**	1982		1992	
1953	**18**	1963		1973	**50**	1983	**18**	1993	
1954	**33**	1964	**50, 18**	1974	**18**	1984	**33**	1994	**33, 18, 33**
1955	**18, 33**	1965	**33, 50**	1975	**18, 50**	1985	**18**	1995	
1956		1966	**18**	1976	**100**	1986		1996	**18**
1957	**18**	1967		1977		1987	**18, 18**	1997	**18**
1958	**33**	1968	**18**	1978	**33**	1988		1998	
1959	**18**	1969		1979	**18**	1989	**33**	1999	**18, 33**

* in thousands of cubic feet per second (1000 ft^3 / sec); some years have more than one flood

1) According to this table, which size flood happens more often?

large small

The table below summarizes the number of times each flood happens in 50 years (data are from the previous table). The recurrence interval indicates how often a flood of that size occurs.

Recurrence interval = 50 years ÷ the number of times that flood occurs in 50 years

2) Determine the recurrence interval (average number of years between floods) for the largest and smallest flood heights. The other recurrence intervals have been calculated for you.

Flood discharge	# of times in 50 Years	Recurrence interval (1 flood every _____ years)
100,000 ft^3/sec	1	
50,000 ft^3/sec	6	8.3
33,000 ft^3/sec	13	3.8
18,000 ft^3/sec	25	

3) Does a flood with a large recurrence interval occur more or less often? more less

Flood Frequency Curve

Below is the flood-frequency curve for this stream. The recurrence interval is plotted compared to the discharge (size of the flood).

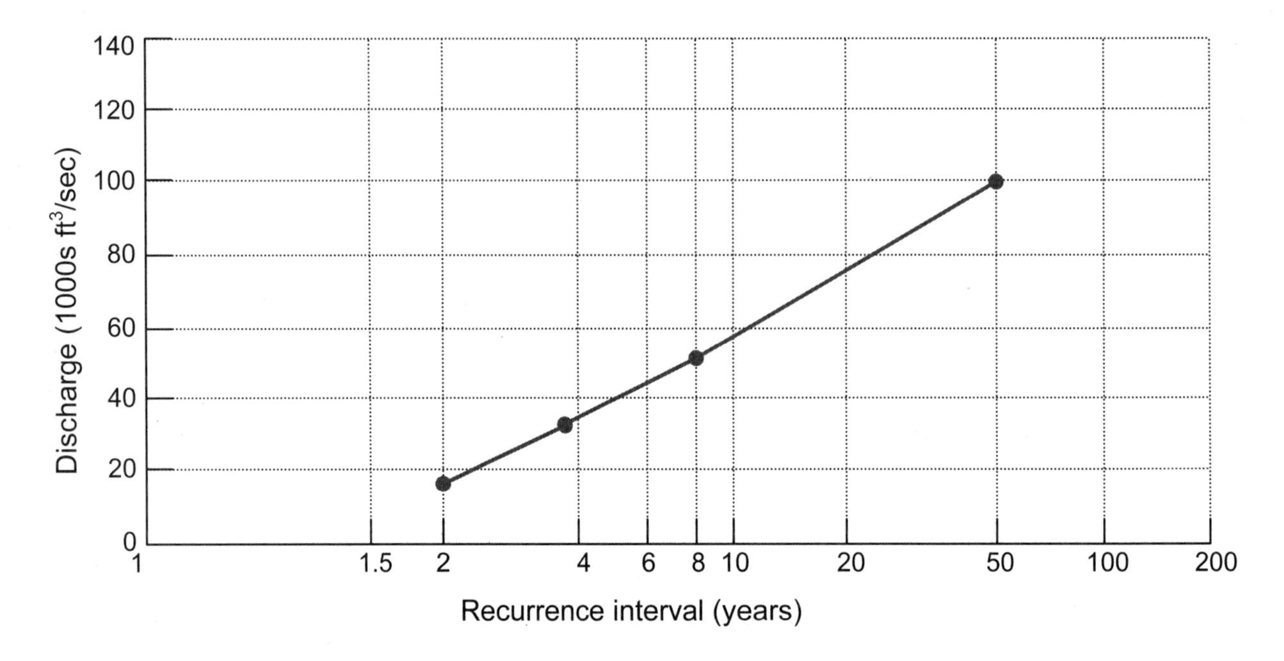

4) Large floods that are 75,000 ft^3/sec occur on average once every _______ years.

5) Small floods that are 20,000 ft^3/sec occur on average once every _______ years.

6) Predict the recurrence interval of an extremely large flood with a discharge of 140,000 ft^3/sec.

Part 2: Predicting Floods

7) The chart makes it look like floods of discharge 50,000 ft^3/s occur exactly once every eight years. Look at the first table. Are floods of certain sizes regularly spaced?

8) If there is a flood with a discharge of 50,000 ft^3/s, will that flood happen again in exactly eight years? Explain your answer.

9) A flood has a recurrence interval of four years. This flood occurs in 2006. What is the chance the flood will happen in 2007?

10) Two students are thinking about this question.

Student 1: *I think that it will not happen in 2007. It is just like an earthquake. If an earthquake just happened, then it will take many years for the stress to build up again for another earthquake. Since the flood just happened in 2006, it is not likely to happen the next year.*

Student 2: *But weather this year doesn't care what happened last year. The flood has a 25% (1-in-4) chance of happening each year, so it would have a 25% chance of happening in 2007. Floods are not evenly spaced over the years.*

Student 1: *I still think it will happen in 2010, exactly four years later. The flood frequency is one flood every four years.*

With which student do you agree? Why?

11) Find an example in the chart on the first page that shows that a large flood can happen two years in a row. Write the years you chose below.

12) A very large flood occurs in a small town. The local tourist board posts the message: "Come visit our city! We'll be safe from floods for another 100 years!" Do you agree with this message? What is the likelihood that the same flood will happen again next year?

The contamination within groundwater flows along with the groundwater. As a result, by determining the flow of groundwater, you can determine the flow of the contamination.

1) The contours on the map to the right show the water table elevation above sea level. Why is Arrow C better than the other arrows at showing the direction groundwater will flow?

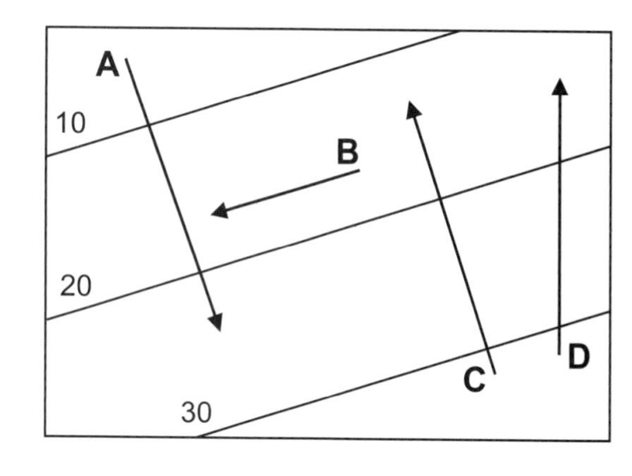

Arrow A:

Arrow B:

Arrow D:

2) Which arrow shows the direction the contamination in the groundwater will flow?

Arrow A Arrow B Arrow C Arrow D

3) The contours on the map to the right show the water table elevation above sea level. The black triangle represents a septic tank. If this septic tank is improperly installed, which location would detect contamination first?

E F G

Explain your answer.

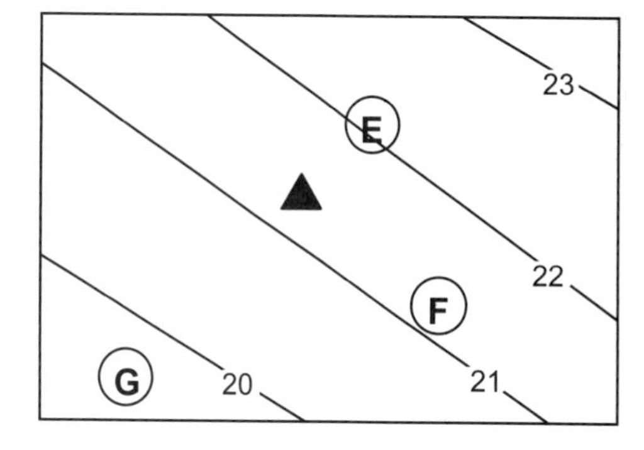

4) Three students are debating which location will first detect the contamination.

Student 1: *I think that E will detect the contamination first because it is closest to the septic tank.*

Student 2: *But you need to consider what direction the water will flow. I think that F will first detect the contamination because the contamination will flow along the lines down toward F.*

Student 3: *The lowest elevation is toward G where the elevation is less than 20, so the water will flow from 21 to 20, and G will detect contamination first.*

With which student do you agree? Why?

Groundwater Contamination

The figure below shows the water table elevation, three houses, three wells, and a septic tank.

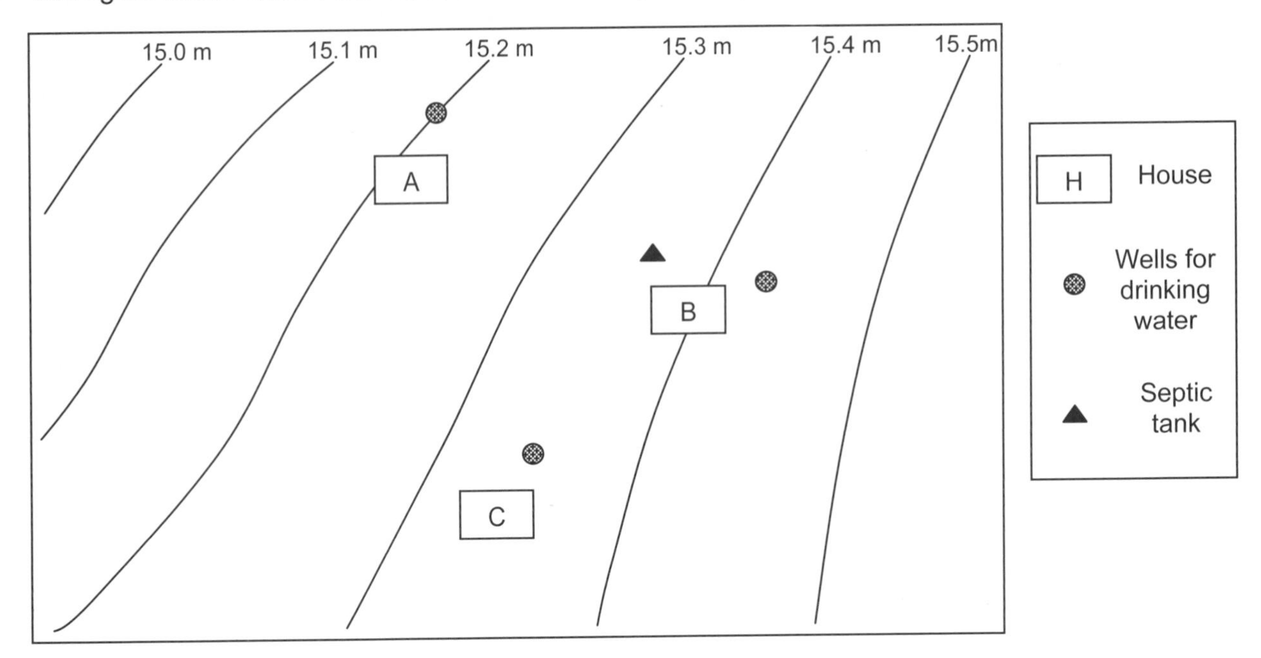

5) Draw at least three arrows on the map showing the direction of ground water movement. Keep in mind your answer to Question 1 when drawing the arrows.

6) The triangle is a proposed location of septic tank for House B. Septic tanks may release contaminants in the water if constructed without the proper considerations. Draw an arrow showing the direction of the contaminant plume which may be released from the septic tank.

7) The wells of which houses, if any, could be affected by the septic tank? A B C

8) Put a star on the map where you think would be a better place for a septic tank for House B. The tank, to be useful to the inhabitants of the house, needs to be located near the house. Explain your decision in the space below.

Part 1: Carbon Dioxide in the Atmosphere

Compare the chart to the right showing the amounts of different elements in the atmosphere with the graphs below showing the carbon dioxide concentrations and temperature anomalies over the last 250 years.

Current Atmosphere Composition of Select Elements	
Nitrogen	78.1%
Oxygen	20.9%
Carbon dioxide	0.038%

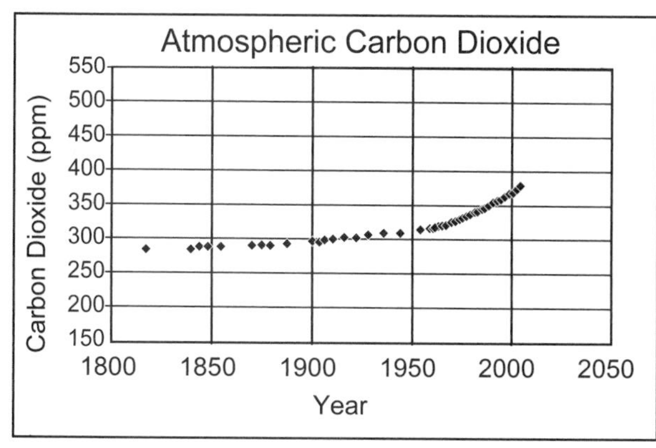

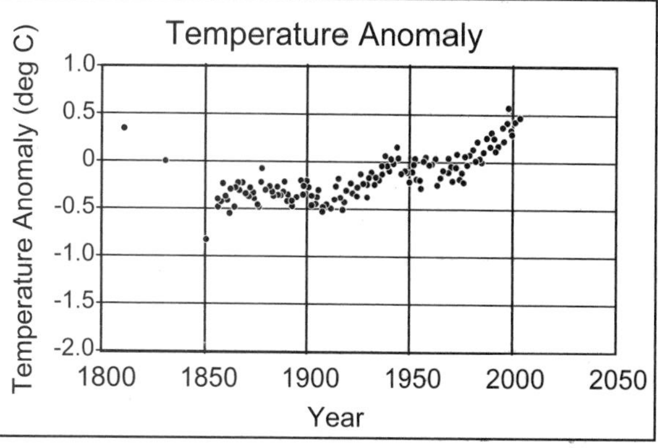

Data from the Carbon Dioxide Information Analysis Center

1) Look at the left graph. What is the carbon dioxide concentration today? __________ ppm

2) What is the trend in carbon dioxide concentration? increasing decreasing constant

Use the trend in the graph to estimate the concentration of carbon dioxide in the atmosphere in 10 years and 100 years. People begin having headaches when carbon dioxide levels reach around 0.5% (5,000 ppm) and lose consciousness when levels reach 10% (100,000 ppm).

3) Will people be able to breathe in 10 years? Yes No

4) Will people be able to breathe in 100 years? Yes No

A temperature anomaly is how much the temperature is warmer or colder than normal. Note that it is measured in degrees Celsius.

5) Look at the graph on the right. What is the temperature anomaly today? __________

6) Therefore, is today warmer or colder than normal? warmer colder

7) What is the trend in temperature anomaly? increasing decreasing constant

8) Based on the graph, what will the temperature anomaly be in 50 years? ___________

9) What is the relationship between carbon dioxide and temperature anomaly?

Climate Change and Carbon Dioxide

Part 2: Comparison of Today to the Past

Below are graphs showing the levels of atmospheric carbon dioxide and the temperature anomaly for the past 400,000 years.

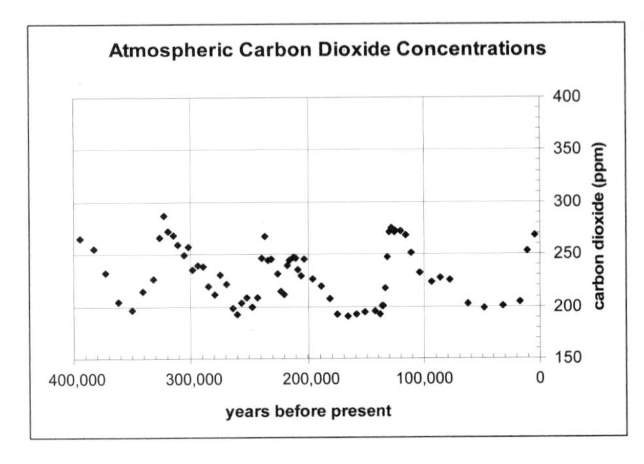

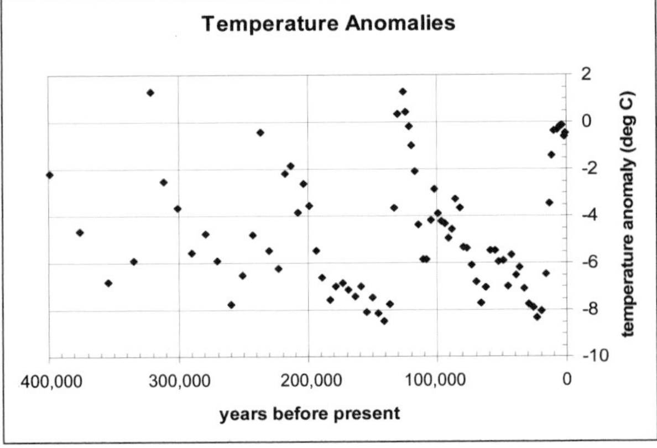

10) Use your answer to Question 1 to plot a point on the left graph indicating the carbon dioxide concentration today (note the different scale). How does the carbon dioxide concentration today compare to the concentrations in the past?

11) Use your answer for Question 5 to plot the current temperature anomaly on the graph above (note the different scale). How does the temperature anomaly today compare to the anomalies in the past?

12) Think about how hot it would need to be before humans could not survive. Did the temperature in the last 400,000 years get so hot that humans could not survive? Yes No

13) Based on past data, do you think the level of carbon dioxide in the atmosphere is going to increase to the point that Earth will warm up so much in the next 500 years that people will not be able to survive? Why or why not?

14) If humans do not need to worry about breathing or heat stroke issues, brainstorm some of the reasons why we are concerned about increasing levels of carbon dioxide in the atmosphere.

Part 1: Earth, Moon, and Sun Relationships

1) How long does it take for Earth to rotate on its axis?

 1 day 1 month 1 year

2) How long does it take for the Moon to travel around Earth?

 1 day 1 month 1 year

3) How long does it take for Earth and the Moon to travel around the Sun?

 1 day 1 month 1 year

4) Why does the Moon appear to travel across the sky in 12 hours (just as the Sun travels across the sky in 12 hours)?

High tides occur on Earth on the sides closest to the Moon and farthest from the Moon. Below is a figure showing Earth and the Moon as viewed from the North Pole. The dashed line on the diagram below shows tides on Earth as a result of the Moon (not to scale).

5) Label the high tides and low tides on the diagram.

6) At any point in time, how many high tides are on Earth?

Moon

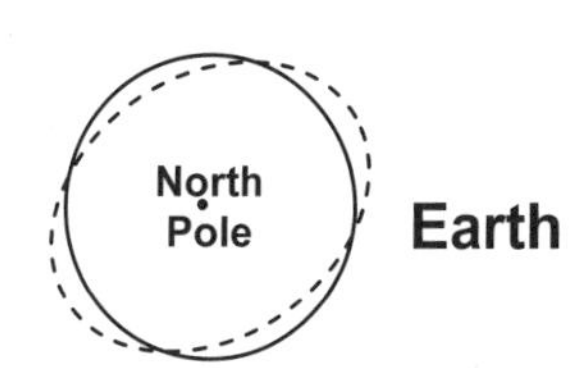

7) Two students are debating about why the tide goes in and out, alternating between high and low tide twice a day.

Student 1: *I think that the tides change each day because Earth spins on its axis beneath the Moon, so a point will go from high to low tide as Earth rotates.*

Student 2: *No, I think the tides change each day because the Moon moves around Earth. As the Moon moves around Earth, the high tides and the low tides follow the Moon.*

With which student do you agree? Why?

Part 2: Semidiurnal Tide Cycle

Semidiurnal tides are two high tides and two low tides in approximately 24 hours.

Below are a set of figures showing Earth and the Moon over 24 hours as viewed from the North Pole. The star represents one place on the equator of Earth. The diagram is not to scale.

8) Draw the tidal highs (bulges) and lows on Earth created by the Moon.

9) Indicate if the star on the surface of Earth has a high tide or low tide at each point in time.

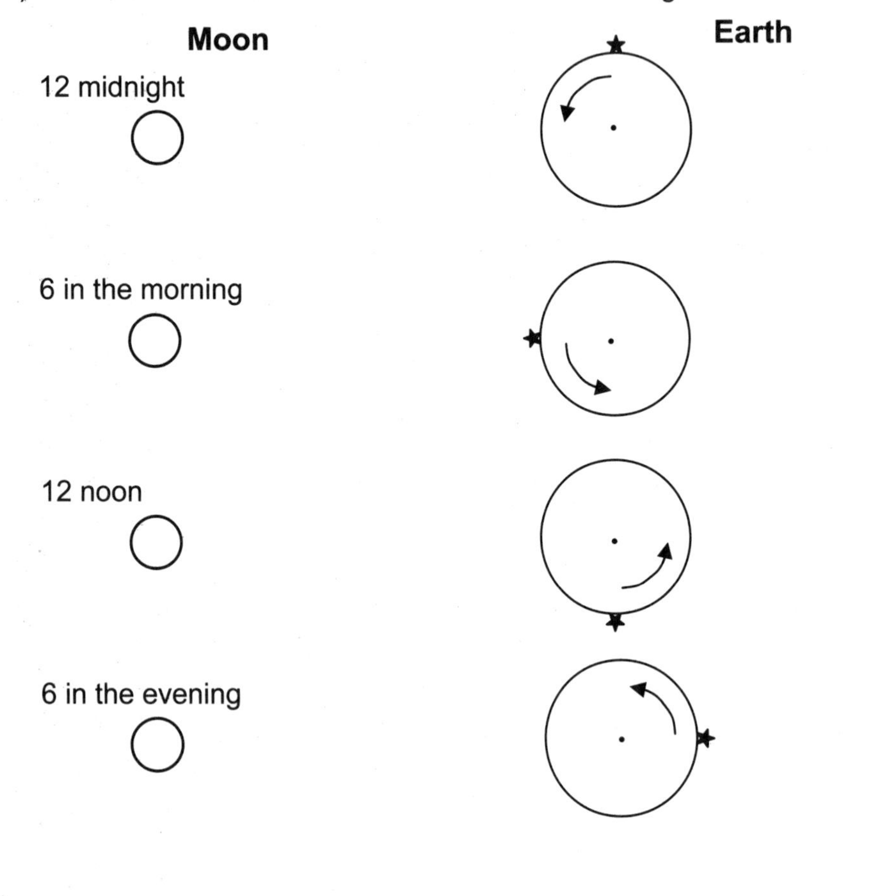

10) How many hours separate a high tide from a low tide? ____________

11) How many hours apart are high tides? ____________

12) Planet Z is discovered. Its day is 10 hours long. Its moon revolves around it in 200 hours. How many hours apart are high tides on Planet Z? Explain how you determined your answer.

13) Summarize (in words or in a clearly labeled diagram) why semidiurnal tides occur and why there is a 24-hour cycle on Earth.

Part 1: Earth, Moon, and Sun Relationships
It takes approximately four weeks (one month) for the Moon to travel around Earth. Below is a diagram showing the Earth, Sun, and Moon as viewed from the North Pole. The diagram is not to scale.

1) Draw an arrow on Earth to show Earth's rotation. How long does this take?

2) Draw an arrow on the Moon showing the direction of the Moon's motion around Earth. How long does this take?

Half of the Moon is shaded, and the other half is bright, showing the side that is lit from the Sun.

3) Circle the diagram to the right that shows what the Moon looks like as viewed from Earth.

4) What phase of the Moon is this? full moon quarter moon new moon

5) Draw in the tides. With a thin line, draw the tides on Earth caused by the Moon. With a dashed line, draw the tides on Earth caused by the Sun. Remember that the tides caused by the Sun are smaller than those caused by the Moon.

Part 2: Spring and Neap Tides
If the high tides from the Sun and Moon are in the same location, they add up, making them extreme (spring tides). If they are in different locations, they subtract from each other (neap tides).

6) In the configuration above, are the tides on Earth extreme or not extreme?

7) What is the name of this kind of tide?

Spring and Neap Tides

8) Draw an Earth-Sun-Moon alignment below during a spring tide. There is more than one correct answer; however, there are incorrect answers. Remember that the Moon is always closer to Earth than the Sun ever is.

Shade half of the Moon, showing which side is lit by the Sun.

9) Circle the diagram to the right that shows what the Moon looks like as viewed from Earth.

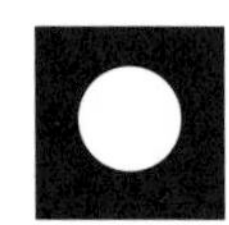  

10) What phase of the Moon is this? full moon quarter moon new moon

11) During what two phases of the Moon do spring tides occur?

full moon quarter moon new moon

12) How long would you be able to see each phase? one day one week one month

13) How many spring tides occur every time the Moon travels around Earth once?

14) Planet Z is discovered. Its day is 10 hours long. Its moon revolves around it in 200 hours. How many hours apart are spring tides on Planet Z? Explain how you determined your answer.

15) Explain why during some weeks tides are higher than during other weeks over the course of a month.

16) During a spring tide, explain why there is a pattern of two high tides and two low tides each day.

Please keep in mind that Question 16 is not asking why spring tides occur, so your answer does not need to mention the tides caused by the Sun.

Part 1: Erosion

Not every place on Earth is a location where rocks are being deposited. Some places are where rocks are being eroded. The rock record is the rocks that remain that help us determine the history of an area.

1) In which of the following environments are rocks most likely to be eroded? This environment is an area where sediments are NOT deposited. Explain your reasoning.
   a. mouth of river
   b. ocean floor
   c. mountain
   d. coral reef

2) How could we tell if erosion happened in the rock record?

3) Two students are debating how erosion would be recorded in the rock record.

Student 1: *If there was erosion, you wouldn't be able to see it because it gets rid of rocks. So there wouldn't be any record of it.*

Student 2: *But, because rocks that were once there were taken away, you would see a missing surface at the top of the rocks. That missing surface would be visible in the rock record.*

With which student do you agree? Why?

4) In the cross section to the right, what two things happened between the deposition of Sediment A and Sediment D?

   1.

   2.

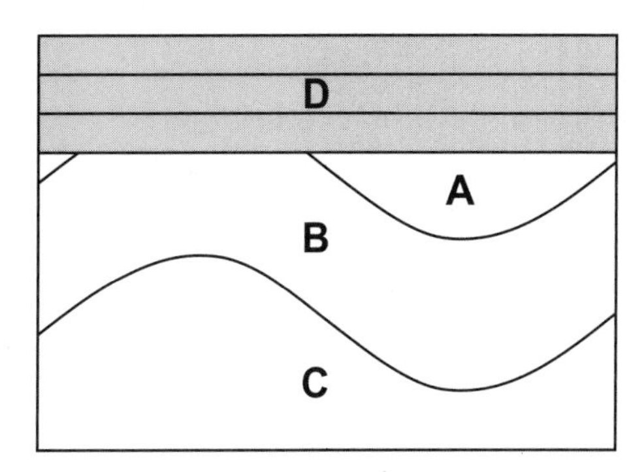

Part 2: Unconformities
In the cross section in Question 4, the surface between A/B and D is called an unconformity. An unconformity in the rock record is typically a surface where there was erosion of the rocks.

5) In which of the following environments is an unconformity most likely to form? Explain your reasoning.
   a. mouth of river
   b. ocean floor
   c. mountain
   d. coral reef

6) Why should your answers to Question 1 and Question 5 match? Change your answers as necessary.

Use the cross section to the right to answer Questions 7–9.

7) Draw a line along the unconformity in the cross section.

8) How do you know there is an unconformity in the cross section?

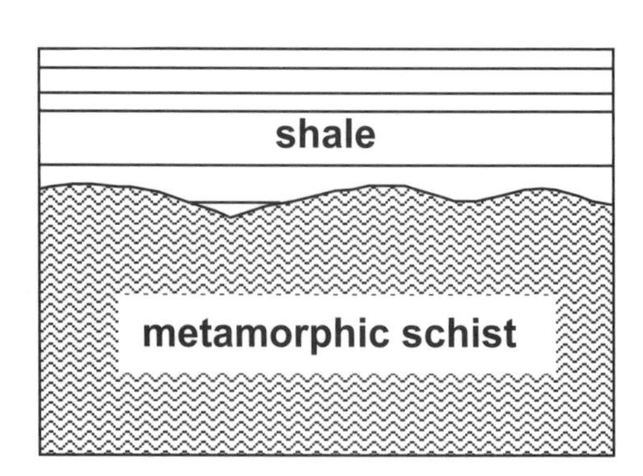

9) Describe three things that happened in the area to form the rocks in the cross section.

10) A person says to you, “I don’t believe that Earth is thousands of millions of years old because there should be more sedimentary rock layers if Earth is that old.” What would you say to that person to tell them that their argument does not make sense?

Part 1: Absolute Age Dating
Absolute ages are ages in years before present (e.g., This rock formed 100 million years ago (100 Ma)). Absolute ages are determined by studying the decay of unstable elements and knowing their half-lives. The table below summarizes characteristics of different unstable elements used to determine absolute geologic ages.

Unstable element system	Used for	Age limits
$^{14}C \rightarrow {}^{14}N$ (carbon-14)	Things that were once alive that retain their original carbon molecules	100 to 50,000 yrs before present (0.05 Ma)
$^{40}K \rightarrow {}^{40}Ar$ (potassium-argon)	Volcanic rocks	0.05 to 4600 Ma
$^{238}U \rightarrow {}^{206}Pb$; $^{235}U \rightarrow {}^{207}Pb$ (uranium-lead)	Volcanic rocks	10 to 4600 Ma

1) What system, if any, can be used to determine the absolute age of dinosaur bones?

2) Three students are debating how to date dinosaur bones.

Student 1: *We can use carbon-14 because it's used to date things that were once alive. The flesh of a dinosaur was composed of carbon.*

Student 2: *But dinosaurs lived more than 65 million years ago, so carbon-14 doesn't work. To date something that old, you need to use potassium-argon or uranium-lead.*

Student 3: *Carbon-14 can't date things as old as dinosaurs and potassium-argon and uranium-lead only work for volcanic rocks, so none of these work.*

With which student do you agree? Why?

Part 2: Relative Age Dating
Relative dating gives the ages of rocks relative to the ages of other rocks (e.g., This rock is older than that rock or fossil). In undeformed sedimentary rocks, the oldest rock layer is found on the bottom and the youngest rock layer is found on the top.

3) The figure to the right is a cross section through sedimentary rock layers. Which layer is the oldest?

A B C D E

4) How old (relative to the other layers) is the dinosaur bone?

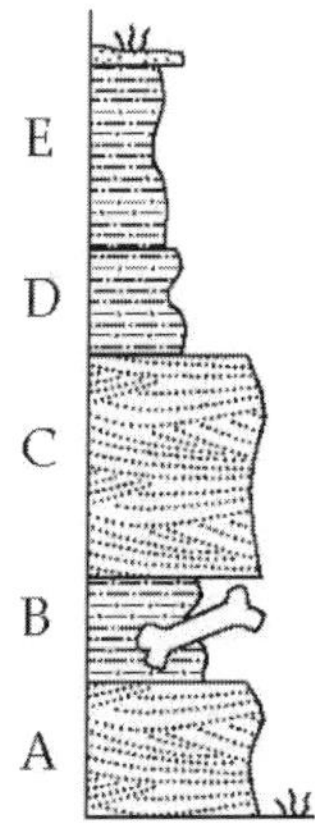

Part 3: Combining Dating Techniques

To the right is a cross section through sedimentary and volcanic rock layers. Sedimentary rock layer B contains a dinosaur bone.

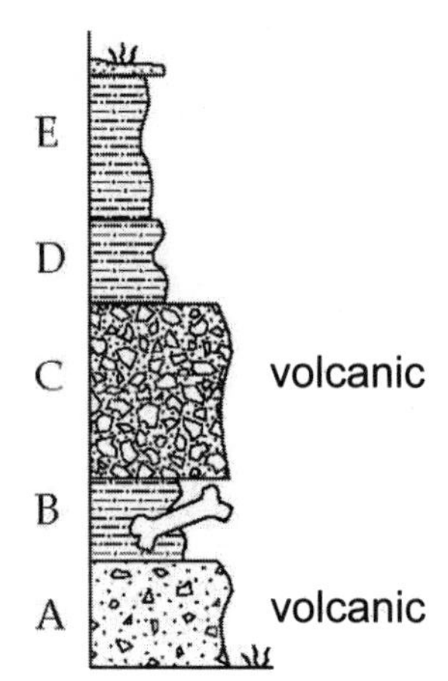

5) Which two of the absolute age dating methods would you use to determine the age of volcanic rock layers A and C?

6) How could you determine the age of the dinosaur bone?

7) If volcanic rock A is determined to be 90 Ma, and volcanic rock C is determined to be 80 Ma, what is the age of the dinosaur bone?

8) Two students are debating the age of the dinosaur bone.

Student 1: *The bone is between 80 and 90 million years old. You can't be more specific because you can't determine how fast the layers were deposited.*

Student 2: *I think the dinosaur lived 85 million years ago because the bone is exactly in the middle of the two volcanic layers.*

With which student do you agree? Why?

Part 1: Using the Principles of Relative Dating
For each cross-section diagram, you will number the units and events in order of formation as directed in the questions. The oldest unit is number 1; continue until the youngest unit is numbered. If there is an unconformity (erosion surface) or faulting, number it as an event.

1)

Conglomerate
Sandstone
Shale
Limestone

a. Number the four events form oldest to youngest. Remember, 1 is the oldest.

2)

A
B
C
C
D
E
F

a. Which is older, Unit A or Unit F?

b. Which is older, Unit C or the erosion surface (shown by the wavy line)?

c. Number the seven events from oldest to youngest.

d. What might have caused the dip where the arrow is pointing?

3)

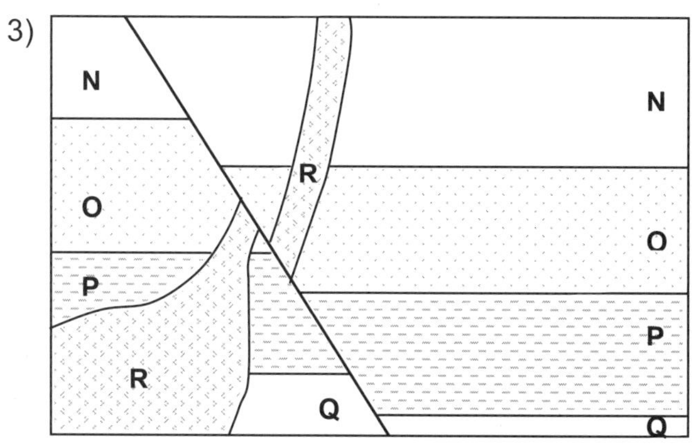

a. What is the diagonal straight line through the cross section? _______________

b. Which unit is a pluton (intrusive igneous rock)? _____________

c. Which is older, Unit R or the fault that cuts Unit R? __________________________

d. Number the six events from oldest to youngest.

Determining Relative Rock Ages

Part 2: Correlating Rock Layers
The two cross sections below show rocks in Arizona and Utah. Fossils are indicated by the symbols in each section.

4) Determine the history of each area and fill in the geologic columns by writing out the sediment type and episodes of erosion (unconformities). Each line will be used once.

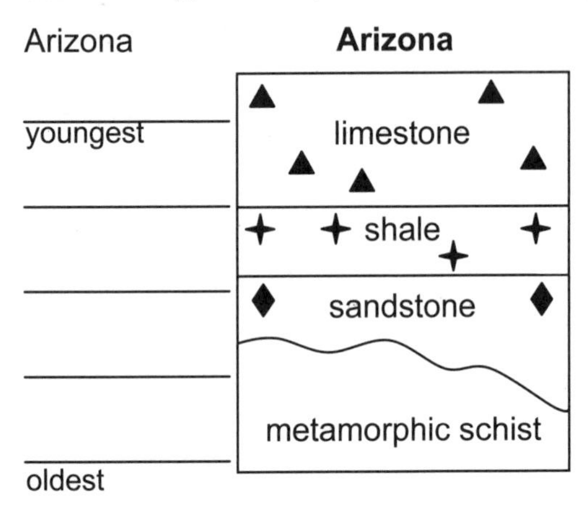

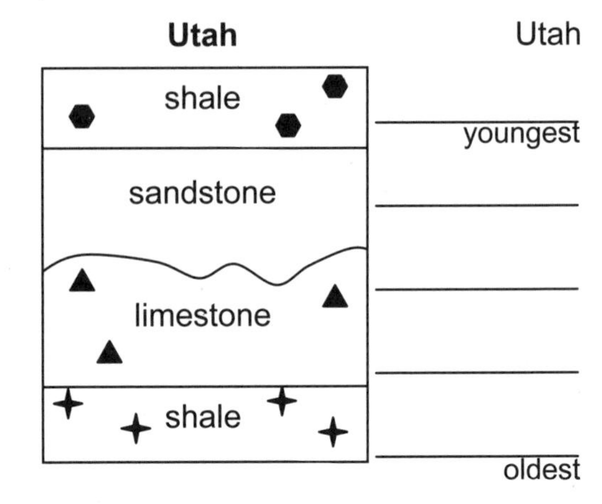

5) Identify the units that have the same fossil in Arizona and Utah.

Draw the fossils here:

6) Draw arrows on the diagram between the Arizona and Utah columns indicating which units are the same age based on fossil evidence.

7) If you went to Utah, what type of rock would you expect to find below the lower shale unit with the ✦ fossil?

8) Determine the history of the entire area by correlating the two cross sections using the fossils present in the units. If there are units in different places with the same fossil, they were deposited at the same time and will be written only once in the combined column.

youngest

____

____

____

____

____

____

____

____

oldest

Part 1: A Magic Half-Life Candy Shop
A candy shop puts 1000 magic green candies in a jar. After 10 years, half of the green candies turn red. After another 10 years, half of the remaining green candies turn red. This process continues until all the green candies turn red.

1) After 10 years, how many candies remain green? _________ How many are red? ________

2) Fill in the table below and graph the results on the chart.

# of years	# of green candies
0	1000
10	
20	
30	
40	
50	

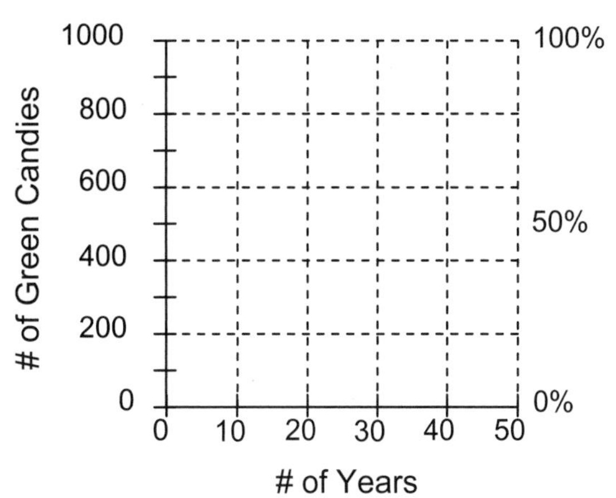

3) When you go to the shop and look in the magic candy jar, 125 candies are green. How many years have the candies been in the jar? Explain how you know, or write down any equations you used.

4) How many years does it take for 75% of the candies to turn red (25% of the candies remain green)?

The half-life of the magic candies is how long it takes for half of them to change colors.

5) What is the half-life of the candies in this jar?

6) In a different candy jar, half of the yellow candies change to blue every 47 years. What is the half-life of the candies in this other jar?

Part 2: Radioactive Elements

Radioactive elements work in a similar way as the candies in the magic candy shop.

7) Match the phrases relating to the magic candy shop to the phrases relating to radioactive elements in minerals. Do this by drawing lines between the two columns.

a. The green candies are put into a jar.

b. All candies begin green.

c. No candies disappear.

d. Candies change from green to red.

e. The half-life of the candy is the time it takes for half the candies to change color.

1. All atoms begin as parent isotopes.

2. Atoms change from the parent to the daughter isotope.

3. The half-life of the radioactive element is the time it takes for half the parent isotope to change to the daughter isotope.

4. No radioactive atoms disappear into nothing.

5. The parent atoms form in a crystal.

To determine the age of a rock (just like determining the time the magic candies have spent in the jar) you need to know two things: 1) how long the half-life is, and 2) how much of the original parent is remaining.

8) Geologists can measure the half-life of radioactive elements in a lab, so we know the half-life for common radioactive elements. Therefore, to determine the age of a particular rock, what would you need to measure in that rock?

A geologist is studying two different rhyolite flows to determine if they were erupted at the same time.

9) Rhyolite #1 has 50% of the Parent Isotope F remaining. How many half-lives have occurred?

10) Rhyolite #2 has 75% Daughter G and 25% Parent H. How many half-lives has the rock gone through?

11) Parent Isotope F in Question 9 has a half-life of 100 million years. Use your answer to Question 9 to determine the age of Rhyolite #1.

12) The half-life of Parent H-Daughter G in Question 10 is 200 million years. Use your answer to Question 10 to determine the age of Rhyolite #2.

13) Are the two rhyolite flows the same age? Yes No

One-hundred kilometers (60 miles) from Rhyolite #1 and Rhyolite #2 the geologist found an ash layer in the rocks. In order to determine how explosive the volcano could be, the geologist wanted to know if the ash came from the same eruption as either of the rhyolites.

14) A different parent-daughter pair in the felsic ash has a half-life of 50 million years. There is 25% of this parent still remaining. How old is the felsic ash?

15) Another parent-daughter pair is measured in the felsic ash. This pair has a half-life of 25 million years. There is 1/16 of the parent remaining. How old is the felsic ash?

16) The felsic ash is the same age as which rhyolite? Rhyolite #1 Rhyolite #2

Part 1: Hypotheses
A scientific hypothesis needs to

1) be supported by the majority of current data

2) be testable

An alien on Earth is wondering why a rubber ball falls back down to the ground after it is thrown into the air. It comes up with several ideas about the ball.

a. Gravity is pulling the ball to the ground.

b. A mystical force that cannot be measured is pushing the ball down.

c. Earth's magnetic field is pulling on the rubber ball.

1) Which statement is NOT a hypothesis because it is not testable? a b c

2) Which statement is NOT a hypothesis because it is not supported by current data? a b c

3) Which statement IS a scientific hypothesis? a b c

Part 2: Dinosaur Extinction
Below are possible scenarios explaining the extinction of the dinosaurs.

a. Dinosaurs were killed off by a virus.

b. A large meteorite impact caused the climate to change so some plants and animals could no longer survive.

c. Volcanic eruptions caused the climate to change so some plants and animals could no longer survive.

d. Mammals ate all the dinosaur eggs.

4) Determine if each statement above is a valid hypothesis. Be sure to explain your answer.

a. Yes No because...

b. Yes No because...

c. Yes No because...

d. Yes No because...

Hypotheses of Dinosaur Extinction

5) Two students are debating the hypotheses of dinosaur extinctions.

Student 1: *I think that the meteorite and volcano statements are valid hypotheses, but the other two are not. You can't test the fossil record to find out if they are true, and they don't explain animals other than dinosaurs going extinct.*

Student 2: *I think that all of the statements are all valid scientific hypotheses explaining dinosaur extinctions. I saw all of them on a dinosaur program on TV, and they all seem possible. No person was there to watch the dinosaur extinction, so all of the scenarios are hypotheses.*

With which student do you agree? Why?